GeoGuide

Series editors

Wolfgang Eder, Munich, Germany
Peter T. Bobrowsky, Burnaby, BC, Canada
Jesús Martínez-Frías, Madrid, Spain
Axel Vollbrecht, Göttingen, Germany

The GeoGuide series publishes travel guide type short monographs focussed on areas and regions of geo-morphological and geological importance including Geoparks, National Parks, World Heritage areas and Geosites. Volumes in this series are produced with the focus on public outreach and provide an introduction to the geological and environmental context of the region followed by in depth and colourful descriptions of each Geosite and its significance. Each volume is supplemented with ecological, cultural and logistical tips and information to allow these beautiful and fascinating regions of the world to be fully enjoyed.

More information about this series at http://www.springer.com/series/11638

Frances M. Williams

Understanding Ethiopia

Geology and Scenery

 Springer

Frances M. Williams
Department of Earth Sciences
University of Adelaide
Adelaide, SA
Australia

ISSN 2364-6497   ISSN 2364-6500 (electronic)
Geoguide
ISBN 978-3-319-02179-9   ISBN 978-3-319-02180-5 (eBook)
DOI 10.1007/978-3-319-02180-5

Library of Congress Control Number: 2015950056

Springer Cham Heidelberg New York Dordrecht London

Printed on acid-free paper

Springer International Publishing AG Switzerland is part of Springer Science+Business Media (www.springer.com)

This book is dedicated to the memory of Dr. Bill Morton, friend and colleague, whose contribution to the understanding of Ethiopia's geology was tragically cut short.

Foreword

To venture is to enlarge one's life. To venture into Ethiopia is to encounter a cornucopia of unmatched physical and cultural diversity. Poised high and commanding above the Nile plains to the west, the Red Sea littoral to the east and the Kenyan savannah to the south, Ethiopia's castellated plateaus have been home, rallying ground and refuge to a two thousand year-long succession of ruling emperors (and empresses!). Ethiopia, today an increasingly powerful influence in Africa, has an ancient name worthy of the birthplace of *Homo sapiens*, expressive of its hospitality to the Queen of Sheba and possibly the Ark of the Covenant, and a refuge for early followers of the prophet Muhammad. Yet the kaleidoscopic human history of Ethiopia has been conditioned by a far older history, that of its landscape. For as the author of this book rightly remarks, "In Ethiopia geology lies behind almost every experience".

The spectacular landscape of Ethiopia has been moulded and sculpted by forces inside the earth which continue to act today: volcanoes simmer in the lowlands, earthquakes episodically open chasmic fissures in Afar, and the Rift Valley continues slowly but inexorably to widen in response to ongoing continental drift. At the same time, great rivers impressed on the highlands continue to roar and cut deep into their awesome canyons: the Abay (Blue Nile), Tekeze, Omo and Shebeli. This is nature alive on the grand scale!

Frances Williams' love affair with Ethiopia stems from close to half a century's acquaintance with the land and its people. She has travelled it from east to west, north to south and peak to gulf. As a field scientist, photographer, linguistic artist, and convivial and sensitive companion, she now expresses in this gem of a book her knowledge of not only the geological tapestry of Ethiopia, but also its archaeology and ancient human cultures. With clarity and, in the best sense, simplicity, she communicates the drama of the moulding of Ethiopia's geology, both for the novice and for those advanced but still learning!

Between these covers, the traveller, on the move or at home, can savour and more fully appreciate the remarkable and complex choreography of Ethiopia's geological evolution. The first section of the book, after introducing some essential geological concepts, describes its 600 million year-long story: from an initial, crunching continental collision; then successively through glaciation near the South Pole, gentle flooding by tropical seas, violent flooding by gigantic white-hot lava flows; and lastly and still ongoing, the uplifting of the high plateaus while at the same time they are being rifted apart. The second section devotes fifteen chapters to the various geological regions of Ethiopia, well illustrated with colourful maps, informative diagrams and photographs that display an artist's eye.

For the artistry of geology teaches us that ultimately "It is the beginning and the end that shadows the civilisations that pass upon the surface".

Paul Mohr
Author of a 1962 "Geology of Ethiopia"

Acknowledgments

So many people have helped me while preparing this book that it is hard to know where to begin expressing my thanks. I will therefore do so where my association with Ethiopia began, with the Department of Geology (now the Department of Earth Sciences) at Addis Ababa University. Thank you to all its staff and students, past and present, for giving me the opportunity to work in your department all those years ago, for opening my eyes to your wonderful country and for continuing to give me encouragement and support throughout this project.

During my trips to Ethiopia over the past ten years, my travel has been arranged by Mr. Tony Hickey and his team at Ethiopian Quadrants. Thank you to Tony and all the Quadrants staff, especially to W/o Manalebsh Tilahun and W/o Etagegnehu Buli for organising unconventional itineraries, and to everybody for being smiling and welcoming whenever I come to your office. My very special thanks go to long-distance drivers Ato Mesfin Debrework, Ato Fuad Ahmed, Ato Kassahun Ejeta and Ato Kassaye Yohannes who never demurred about going off the beaten track to follow the strange whims of a geologist and for whose driving skills I have unceasing admiration.

Ato Bedassa Gubena kindly arranged logistics for trips to Lake Tana, and to the Semien Mountains where Ato Yalew Tafete was an indefatigable and knowledgeable guide. Chief Warden Ato Maru Biadglegn and his team at the Semien Mountains National Park headquarters at Debarek provided valuable logistical support.

Dr. Yirmud Deneke and the staff of the Awash Falls Lodge have always given me a warm welcome and provided assistance during my visits to Awash National Park.

Many chapters of the book have benefitted enormously from careful review and critical comments by people who are more expert in particular areas than I am. My heartfelt thanks go to Professor Ernesto Abbate, Ato Tadesse Alemu, Professor Mohamed Abdelsalam, Emeritus Professor Michael Beyth, Ms. Morgan Blades,

Dr. Tadewos Chernet, Professor Alan Collins, Professor Giacomo Corti, Associate Professor Derek Keir, Professor Henry Lamb, Professor Paul Mohr, Mr. Robert Neil Munro, Professor Dr. Jan Nyssen, Mr. Peter Purcell, Mr. Keith Rawolle, Mrs. Jan Rawolle, Professor Solomon Tadesse and Dr. Jacques Varet for taking time to read chapters and make constructive comments—and for saving me from some potentially embarrassing errors. I am alone responsible for any errors which remain.

Professor Ernesto Abbate, Dr. Zelalem Assefa, Ms. Morgan Blades, Mr. Andrew Dakin, Mr. Emile Farhi, Dr. Lorraine Field, Professor John Foden, Mr. Eugenio Lizardi, Mr. Robert Neil Munro, Mr. Peter Purcell and Ato Kassaye Yohannes kindly permitted me to use photographs which they had taken. Professor Allan Pring of the South Australian Museum provided the photograph of Ethiopian opal and Dr. Deborah Haynes photographed the diatoms for Chap. 17.

Professor Ernesto Abbate, Professor Mohamed Abdelsalam, Ms. Morgan Blades, Dr. Seifu Kebede, Professor Henry Lamb, Dr. Daniel Mège and Mr. Peter Purcell allowed me to use information from unpublished reports or gave me access to pre-publication material, and Dr. Atalay Ayele of the Institute for Geophysics and Space Research, Addis Ababa, provided the unpublished earthquake map of the Horn of Africa. Mr. Emile Farhi, Dr. Thomas Wanner, Dr. David Haberlah, Ms. Jutta Von dem Bossche and Dr. Ernst Hirsch kindly translated Italian and German texts into English for me. Dr. Yohannes Lemma Didana introduced me to GeoMapApp and instructed me in its use.

Professor Solomon Tadesse arranged my visits to the mines at Adola and Kenticha. My thanks go to him, to Ato Million Mered, Dr. Zerihun Deseta and Ato Abdela Kedir for permitting me to visit, and to Ato Ebisa Dugasa, W/t Meseret Seife, Ato Amsali Yizengau and Ato Mulugeta Abraham for showing me around the mining operations.

The Geological Survey of Ethiopia has provided assistance and support throughout the project, and a great deal of material in the form of maps and reports. I especially appreciate them for providing me with a complete coverage of the 1:250,000 geological maps of Ethiopia and the accompanying notes, which have given a wealth of information. W/o Tenaye Hailu and Ato Girma Tefera patiently downloaded all this material and spent considerable time searching out other references that I required. Ato Tadesse Alemu and Dr. Girma Woldetensae provided valuable information and useful guidance, particularly for my trip to western Ethiopia.

My warm thanks go to the people of Ethiopia, for their friendliness and courtesy, and for unfailingly offering assistance, guidance and hospitality during my travels around their country.

The Department of Earth Sciences, University of Adelaide, granted me an honorary position and provided me with a home in which to work on the final compilation of this book, as well as use of their facilities and their logistical and moral support.

Special thanks go to my husband Emeritus Professor Martin Williams who read through all the chapters and never ceased to give encouragement.

Last but not least, this project would never even have been dreamed of had not Dr. Johanna Schwartz of Springer DE suggested it. Thank you to her, and to Springer, for initiating the project and for bringing it to light.

Digital elevation models for Figs. III.1, 9.1, 12.1, 12.4, 13.1, 15.1, 15.2, 16.3, 19.1, 20.3, 20.6 and 23.1 were obtained using GeoMapApp http://www.geomapapp. org (W.B.F. Ryan, S.M. Carbotte, J.O. Coplan, S. O'Hara, A. Melkonian, R. Arko, R.A. Weissel, V. Ferrini, A. Goodwillie, F. Nitsche, J. Bonczkowski, and R. Zemsky (2009), Global Multi-Resolution Topography synthesis, Geochem. Geophys. Geosyst., 10, Q03014).

Each of the geological maps was constructed using a number of sources, but most are based initially on the 1:2,000,000 Geological Map of Ethiopia (2nd edition) published by the Geological Survey of Ethiopia, with details modified using the regional 1:250,000 geological maps, information from published reports, and my own observations.

Figures 12.2, 17.3, 19.4, 20.4 and PHOTO 19.7 are based on Google Earth imagery.

Contents

Part I   Setting the Scene

1   The Big Picture 3
   1.1   The Earth's Interior 3
   1.2   Moving Plates 5
   1.3   Plate Boundaries 6
   1.4   Where Is Ethiopia in All This? 8

2   Geological Time 11
   2.1   The Geological Time Scale 12

3   Some Notes About Rock Names 15

4   Ethiopia Today 19
   4.1   Ethiopia's Topographic Regions. 19
   4.2   The Geological Map. 21

Part II   Ethiopia's Geological Story

5   Beginnings: The First Half-Billion Years 27

6   The Early and Middle Years: Ice, Sea and Sand 33
   6.1   The Palaeozoic Era: Blank Pages and Icy Spells 34
   6.2   The Mesozoic Era: Sand, Sea and Early Warning Signs 36

7   The Onset of Turbulent Times 39

8   The Break-Up of a Continent, and a Summary of Events   45

Part III   A Journey Through Geology and Scenery

9   The Great Gorges: Slices Through Time   57
     9.1   The Blue Nile Gorge   57
     9.2   The Tekeze Gorge   67

10   Western Ethiopia: Evidence of a Continental Collision   73

11   Landforms, Monuments and Hidden Churches of Tigray   83
     11.1   The Mekele Basin   85
     11.2   Hauzien to Adigrat, and the Hidden Churches
            of Tigray.   87
     11.3   Adua and Axum: Plugs, Domes and Obelisks   95

12   The Western Highlands: Lava Flows and Great Volcanoes   103
     12.1   The Trap Series   106
     12.2   The Shield Volcanoes and the Semien Mountains.   111
     12.3   The Southern Part of the Western Highlands
            and the Dividing Line.   117

13   Lake Tana and the Blue Nile   121
     13.1   Lake Tana.   121
     13.2   The Source of the Blue Nile   128
     13.3   The Blue Nile and the Great Loop   130

14   The Geology and Churches of Lalibela   135

15   The Southeastern Highlands and the Ogaden.   153
     15.1   The Bale Mountains   157
     15.2   The Caves of Sof Omar and Mechara.   161
     15.3   The Valley of Marvels   164
     15.4   The Marda Fault Zone and the Marda Pass   165
     15.5   The Ogaden Plains.   167

16   Introduction to the Rift Valley   171

17   The Rift Valley Southwards: Volcanoes,
     and Lakes Ancient and Modern 179
     17.1   Ignimbrite 182
     17.2   The Rift Volcanoes 183
     17.3   The Rift Lakes. 187
     17.4   The Wonji Fault Belt 193
     17.5   The Amaro Horst. 195

18   South of Lake Chamo: A Transition Region 197
     18.1   "New York" 200
     18.2   Stone Age Peoples 203
     18.3   El Sod: An Explosion Crater or Maar. 203
     18.4   The Adola Belt: Gold and Tantalum. 205

19   The Rift Valley Northwards: Volcanoes, Fissures
     and Fresh Lava Flows 209
     19.1   Volcanoes of the Northern Rift 210
     19.2   K'one Volcano 211
     19.3   Fantale Volcano and Awash National Park 213
     19.4   The Puzzle of Lake Besaka 219
     19.5   The Wonji Fault Belt in the Northern Main
             Ethiopian Rift 220

20   The Rift Margins and the Great Western Escarpment 225
     20.1   The Margins of the Main Ethiopian Rift 227
     20.2   The Rift Margin in the Addis Ababa Region 229
     20.3   The Debre Zeit Craters 230
     20.4   The Rift Margin from Addis Ababa to Lake Chamo. ... 232
     20.5   The Southern Margin of Afar 234
     20.6   The Western Margin of Afar: The Great Escarpment 235

21   The Enigma of Afar 243

22   Northern Afar: The Birth of an Ocean 249
     22.1   The Volcanic Ranges 251
     22.2   The Erta Ale Range 251
     22.3   Erta Ale Volcano. 253
     22.4   An Ancient Sea Bed. 258
     22.5   The Salt Plain 258

22.6   Dallol.   262
22.7   The Other Volcanic Ranges.   264
22.8   The Manda Hararo Range: Rifting in Action   265
22.9   The Danakil Alps.   268
22.10   Other Volcanoes in Northern Afar   269
22.11   How Old Is Afar? The Red Series
        and Granite Intrusions.   269

23   **Southern and Central Afar: Lava Flows**
        and the Birthplace of Mankind   271
23.1   The Stratoid Series.   271
23.2   The Afar Triple Junction.   276
23.3   The Tendaho Graben   277
23.4   Spreading Axes of Central and Southern Afar   278
23.5   The Birthplace of Mankind   279

24   **Earthquakes in Ethiopia**   285
24.1   Serdo, March 1969.   285
24.2   Other Major Earthquakes in Ethiopia   289
24.3   Distribution of Earthquakes in Ethiopia.   291

25   **Putting It All Together**   295
25.1   What Makes a Continent Break Apart?   295
25.2   How Does a Continent Break Apart?   296
25.3   Is the Ethiopian Situation a Typical Example?   298

26   **What Next? Ethiopia in the Future**   301
26.1   The Long Term Future—Some Speculations   301
26.2   The Near Future—Some Apprehension   303

Glossary   305

Bibliography   319

Index of Localities   333

Index of Topics   339

About the Author

Frances M. Williams taught geology at Addis Ababa University from 1969 to 1976. During that time, she travelled throughout Ethiopia, exploring its geology with her students, mapping volcanoes and investigating the effects of earthquakes. Now relocated to Australia, she visits Ethiopia frequently and has authored and co-authored a number of guidebooks and brochures on localities of geological interest.

Figure 17.4   The story of the Galla Lakes   192
Figure 17.5   Graben and horsts.   195
Figure 18.1   Geological map of southern Ethiopia.   198
Figure 19.1   Digital elevation model of the northern section
               of the Main Ethiopian Rift.   211
Figure 19.2   Geological map of the northern Main Ethiopian
               Rift.   212
Figure 19.3   Geological sketch map of Awash National Park   214
Figure 19.4   Expansion of Lake Besaka.   222
Figure 20.1   Rift margin profiles.   226
Figure 20.2   Illustrating how a rift margin fault becomes tilted
               upwards   227
Figure 20.3   Digital elevation model showing the Ethiopian Rift
               Valley margins, and the southern margin of Afar   228
Figure 20.4   Satellite image of Debre Zeit crater lakes.   231
Figure 20.5   Geological cross section the southern margin of Afar
               at Dire Dawa   235
Figure 20.6   Digital elevation model of the western margin
               of Afar   236
Figure 20.7   Geological section across the Garsat Graben   242
Figure 21.1   Geological map of Afar.   244
Figure 21.2   How Afar may have formed.   247
Figure 22.1   The current situation in the Ethiopia region   250
Figure 22.2   Geological map of the Erta Ale Range   252
Figure 22.3   Aerial photo of Erta Ale volcano   255
Figure 23.1   Digital Elevation Model showing part
               of southern Afar.   274
Figure 24.1   Earthquake map of the Ethiopia region   292
Figure 25.1   The main events in the break-up of the
               African-Arabian plate   297
Figure 26.1   The possible situation 10 million years from now   302

List of Photos

PHOTO 6.1	Tillite from northern Ethiopia	35
PHOTO 7.1	Trap Series basalt flows	44
PHOTO 9.1	View over the Blue Nile Gorge	61
PHOTO 9.2	Columnar jointing in Trap Series basalt	62
PHOTO 9.3	Cliff of Antalo Limestone in the Blue Nile Gorge	63
PHOTO 9.4	Fossil brachiopod shells from the Blue Nile Gorge	64
PHOTO 9.5	Goha Tsion Formation, Blue Nile Gorge	65
PHOTO 9.6	Adigrat Sandstone cliffs, Blue Nile Gorge	66
PHOTO 9.7	Fossil tree trunk in Adigrat Sandstone, Blue Nile Gorge	67
PHOTO 9.8	The gorge of the Jamma River near Debre Libanos	68
PHOTO 9.9	The Tekeze River	69
PHOTO 9.10	Slate near Tekeze River crossing	71
PHOTO 10.1	Serpentinite near Tulu Dimtu	76
PHOTO 10.2	Migmatite east of Gimbi	77
PHOTO 10.3	Serpentinised dunite near Daliti	78
PHOTO 10.4	Granite gneiss at the Didessa River	79
PHOTO 10.5	Scenery north of Gimbi	80
PHOTO 10.6	Zircon crystals	81
PHOTO 10.7	Volcanic plugs between Asosa and Kurmuk	82
PHOTO 11.1	Surface karst weathering, Tigray	86
PHOTO 11.2	Scenery to the south of the Wukro to Hauzien road	88

PHOTO 11.3	Precambrian basement and Enticho Sandstone contact near Hauzien	89
PHOTO 11.4	Ascent to Abuna Yemata Guh church	90
PHOTO 11.5	The entrance to Abuna Yemata Guh church	90
PHOTO 11.6	Ceiling of Abuna Yemata Guh church	91
PHOTO 11.7	Carved pillars and cupola of Mikael Kurara church	91
PHOTO 11.8	The Gheralta ridge, southwest of Hauzien	92
PHOTO 11.9	Basalt dyke staircase at Gheralta	93
PHOTO 11.10	Mikael Kurara church	94
PHOTO 11.11	Basalt church bells of Mikael Kurara	95
PHOTO 11.12	Debre Damo amba	96
PHOTO 11.13	Plugs and volcanic hills near Adua	97
PHOTO 11.14	Obelisk at Axum	99
PHOTO 11.15	Quarry on Gobedra Hill, Axum	101
PHOTO 12.1	The mountain fortress of Mekdela	104
PHOTO 12.2	Dyke through basalt flows near Debre Sina	107
PHOTO 12.3	Intertrappean sediments near Chilga, north of Lake Tana	108
PHOTO 12.4	"God's Finger", a volcanic plug north of Gonder	109
PHOTO 12.5	Opal from the Ethiopian highlands	110
PHOTO 12.6	Undulating, fertile land of the Western Highlands near Gonder	110
PHOTO 12.7	Dissected plateau between Gonder and Addis Ababa	111
PHOTO 12.8	Lava flows, Semien Mountains	112
PHOTO 12.9	The Semien escarpment	115
PHOTO 12.10	The Semien escarpment, near Chenek camp	116
PHOTO 12.11	Imet Gogo, Semien Mountains	117
PHOTO 12.12	Wonchi volcano, near Ambo	119
PHOTO 12.13	Layers of white ash, Wonchi volcano	120
PHOTO 13.1	Zen Akwashita lava tunnel, near Injibara	125
PHOTO 13.2	Boulders of Quaternary basalt on the shore of Lake Tana	126
PHOTO 13.3	Dek Island, Lake Tana	127
PHOTO 13.4	Narga Sellassie monastery on Dek Island	128

PHOTO 13.5   The Little Abay at Gish Abay   129
PHOTO 13.6   The Tis Isat Falls   132
PHOTO 14.1   The rock hewn church of Bet Giorgis, Lalibela   136
PHOTO 14.2   View of Lalibela from Asheton Mountain   139
PHOTO 14.3   Close-up of tuff at Bet Giorgis.   140
PHOTO 14.4   A volcanic bomb at Bet Giorgis.   141
PHOTO 14.5   Contact between tuff and dark grey basalt
              at Bet Medhane Alem.   142
PHOTO 14.6   Monitoring crack in the wall of Bet Raphael   143
PHOTO 14.7   Damage by water seepage to the base
              of Bet Emmanuel   144
PHOTO 14.8   Entrance to the monastic church of Asheton
              Mariam.   145
PHOTO 14.9   Fossil wood in volcanic ash near Asheton Mariam   146
PHOTO 14.10   Nakuta La'ab monastery   147
PHOTO 14.11   Drip collectors in Nakuta La'ab   148
PHOTO 14.12   Imrahana Christos church   149
PHOTO 14.13   Decorated ceiling of Imrahana Christos church.   150
PHOTO 14.14   Recent rock-hewn churches near Checheho   151
PHOTO 15.1   Precambrian migmatite by road to Dire Dawa   157
PHOTO 15.2   Part of a moraine ridge in the Bale Mountains   158
PHOTO 15.3   Garba Guracha, a cirque lake in the Bale
              Mountains.   158
PHOTO 15.4   Glacial striations, Bale Mountains   160
PHOTO 15.5   Volcanic plug in the Bale Mountains   160
PHOTO 15.6   Webi Shebeli gorge near Sheik Hussein   162
PHOTO 15.7   The Chamber of Columns in the Sof Omar caves.   163
PHOTO 15.8   Rock art from Dessa Cave, Dire Dawa region   163
PHOTO 15.9   Balancing rock in the Valley of Marvels   165
PHOTO 15.10   The Marda Range, near Jijiga   166
PHOTO 15.11   Basalt dyke, eastern Ogaden   168
PHOTO 15.12   Meanders of the ancestral Webi Shebeli, depicted
              by a flow of basalt.   169
PHOTO 17.1   The flat floor of the Main Ethiopian Rift west
              of Lake Zwai.   180
PHOTO 17.2   Pumiceous ignimbrite near Lake Langano   183
PHOTO 17.3   Alutu volcano, between Lakes Zwai and Langano   185

PHOTO 17.4   Improvised sauna on Chabbi volcano,
              near Shashamane 186
PHOTO 17.5   Obsidian flow on Chabbi volcano 186
PHOTO 17.6   Fault bounding Lake Langano 189
PHOTO 17.7   Lake sediments in the gorge of the Bulbula River 190
PHOTO 17.8   Diatoms from Bulbula Gorge................... 191
PHOTO 17.9   View of Lake Abaya, Lake Chamo
              and the "Bridge of God"...................... 193
PHOTO 17.10   Hills of recent basalt, on the "Bridge of God"
              in Nech Sar National Park.................... 194
PHOTO 17.11   Faults of the Wonji Fault Belt, east of Lake Zwai 195
PHOTO 18.1   Granite inselberg near Turmi 199
PHOTO 18.2   Granite tor near Yabelo...................... 200
PHOTO 18.3   Termite mound near Mega..................... 201
PHOTO 18.4   Sapphire in river gravels, Yabelo area........... 202
PHOTO 18.5   "New York", near Konso 202
PHOTO 18.6   El Sod explosion crater, near Mega 204
PHOTO 18.7   Olivine nodule from the rim deposits of El Sod 205
PHOTO 18.8   Panning for gold in the gravels of the Dawa River.... 206
PHOTO 18.9   Spodumene in pegmatite at Kenticha
              tantalum mine 207
PHOTO 18.10   Hand-sorting tantalite at Kenticha mine........... 208
PHOTO 19.1   Fresh cinder cone in the northern Main
              Ethiopian Rift 210
PHOTO 19.2   Fantale volcano 213
PHOTO 19.3   Close-up of the Fantale welded tuff 215
PHOTO 19.4   Unbroken blister in the Fantale welded tuff 216
PHOTO 19.5   A broken blister in the Fantale welded tuff 217
PHOTO 19.6   Open fissure in the Fantale welded tuff............ 218
PHOTO 19.7   Sabober tuff ring 219
PHOTO 19.8   Hot pool at Filweha, Awash National Park 220
PHOTO 19.9   Awash Falls, Awash National Park............... 221
PHOTO 19.10   The Awash gorge, Awash National Park 222
PHOTO 19.11   The end of the road at Lake Besaka 223
PHOTO 20.1   Mt Wachacha, near Addis Ababa 229
PHOTO 20.2   Crater rim deposits at Lake Aranguadi, Debre Zeit 232
PHOTO 20.3   Stelae at Tiya 233
PHOTO 20.4   Ara Shatan maar, "The Devil's Lake",
              near Butajira 234

PHOTO 20.5   View through the "Afar Window" near Debre Sina. . . .   237
PHOTO 20.6   The Robit marginal graben   238
PHOTO 20.7   The road to Lake Ashengi.   239
PHOTO 20.8   Lake Ashengi   240
PHOTO 20.9   Amba Alaji   241
PHOTO 20.10   The Belekiya Mountains at the foot
              of the western escarpment   242
PHOTO 22.1   Track to Erta Ale volcano, northern Afar   253
PHOTO 22.2   Ropey lava, Erta Ale   254
PHOTO 22.3   Erta Ale summit caldera   255
PHOTO 22.4   Erta Ale lava lake   256
PHOTO 22.5   Lava fountain at night.   257
PHOTO 22.6   The Salt Plain   259
PHOTO 22.7   Salt and mud crust, Salt Plain   260
PHOTO 22.8   Salt-mining on the Salt Plain near Dallol.   261
PHOTO 22.9   Salt cutter near Dallol.   261
PHOTO 22.10   Camel train and Western Escarpment foothills   262
PHOTO 22.11   Salt mound in Dallol caldera   263
PHOTO 22.12   Green lake at Dallol   264
PHOTO 22.13   Iron-stained salt crystals at Dallol.   265
PHOTO 22.14   Salt "mushroom" at Dallol.   266
PHOTO 22.15   Salt castles at Dallol.   266
PHOTO 22.16   The fissure-vent at Dabbahu, September 2005   267
PHOTO 22.17   The Red Series, at the foot of the Western
              Escarpment   270
PHOTO 23.1   Fault scarp near Mille.   272
PHOTO 23.2   A common southern Afar scene   273
PHOTO 23.3   Dobi Graben   275
PHOTO 23.4   Ayelu volcano near Gewani.   276
PHOTO 23.5   Faults of different trends, between Mille and Logia . . .   277
PHOTO 23.6   Beds of lake sediments and volcanic ash at Hadar   279
PHOTO 23.7   "Lucy", shortly after her discovery in 1974   280
PHOTO 24.1   Aftermath of the 1969 Serdo earthquake   286
PHOTO 24.2   Serdo house, after the earthquake   287
PHOTO 24.3   Serdo school, after the earthquake   288
PHOTO 24.4   Serdo's brick water tower   289
PHOTO 24.5   Dislodged debris near Karakore   290

List of Tables

Table 3.1   Summary of the main igneous rocks encountered
             in this book   16
Table 23.1   The main palaeo-anthropological sites in the middle
             Awash region.   282

Introduction

I first fell in love with Ethiopia before even setting foot in the country. It was early morning on a September day in 1968, and I was on the last leg of my journey to a new job in a country about which I knew almost nothing. As I gazed out of the plane window, I was mesmerised! Little round houses clustered into tiny villages, with wisps of smoke rising from their roofs, a peaceful landscape of patchwork fields that plunged without warning into vast gorges with silvery threads of river meandering in their depths, a sudden steep mountain range and—joy of joys—a beautiful, perfectly circular volcano with a deep blue lake nestling in its crater. I knew at once that this was going to be a very special place for me.

And so it has remained, though there have been many changes since those early days. At the time of writing this book, 2015 in the Gregorian calendar and 2007 in the Ethiopian, the rate of change in Ethiopia is unprecedented. New roads are being constructed, new hotels and lodges opening, new dams and industries being built, and every city and town is throwing its tentacles into the surrounding countryside. Places that were almost inaccessible when I first lived in Ethiopia can now be reached by tarmac road. Comfortable, even luxurious, hotels and lodges have sprung up in places where camping used to be the only option. Progress is both inevitable and desirable but, if allowed to proceed unchecked, may come at a high price. As a geologist, I am keenly aware that there is a very real danger of Ethiopia's important and sometimes unique geological features being overlooked and destroyed in the rush to develop. One of the aims of this book is to bring to the attention of the people of Ethiopia, to visitors to the country and to those responsible for determining her future development, the value of her geological features and heritage. Once lost, these can never be restored.

My primary aim, however, is to share with the reader the thrill of understanding something of Ethiopia's geology. A knowledge of geology, even if only a smattering, will surely enhance the experience of anyone who travels in the country, be they visitor, resident or those who travel vicariously via armchair. In Ethiopia, geology lies behind almost every experience. Her spectacular scenery is determined by geological processes. Geology has made possible the construction of the wonderful rock churches of Lalibela, provided niches for the hidden churches of Tigray and supplied building material for the magnificent obelisks of Axum. Her wildly varying climate, from the fresh coolness of the highlands to the fierce heat of the Afar desert, her lakes and rivers where travellers may enjoy a relaxing swim or the excitement of white water rafting, and the mountains and canyons which make every journey in Ethiopia an adventure, are all products of her geology.

Geology has also played its part in the history of Ethiopia, providing her with a formidably mountainous interior, bounded by great escarpments and faced by inhospitable desert, which have assisted her fiercely independent and resourceful inhabitants to discourage invaders and enable her to be the only country in Africa that was not colonised by a European power.

This book does not set out to be a comprehensive treatise on the geology of Ethiopia. To produce such a work would be a marathon task requiring several volumes. There is a vast amount of specialist literature available to those who wish to delve into details. Rather, it attempts to convey in non-technical language a sense of wonder, founded upon a basic understanding. It is hoped that both geologists and non-geologists will find the book informative and enjoyable. I have tried to assume no prior geological knowledge on the part of the reader and to this end have restricted the use of technical geological terms as far as possible. Since it is impossible to avoid them entirely, I have provided an explanation wherever such terms are used in the text and a glossary to explain them at the end of the book.

Neither is the book intended as a tourist guide. Even if it were, change is too rapid for this to be practicable. New roads appear and existing ones are re-routed. Places once accessible become so no longer as private developments engulf them, for example the shores of the crater lakes at Debre Zeit. The reverse occurs as new roads are constructed, as is the case in northern Afar. Potential landmarks such as hotels appear and disappear. Even the geology itself is not static, particularly in a land as geologically active as Ethiopia. Fissures appear; volcanoes erupt; hotsprings and fumaroles come and go; and lakes may shrink, grow or disappear altogether. Although the book describes many specific localities of geological interest, the above reasons render it difficult to give detailed routes and directions to them. The maps shown in the book are intended as sketch maps only and not for navigation. A reasonably good map of the country will enable you to locate most of the places mentioned, and Ethiopia is a land of helpful and friendly people, and knowledgeable guides, who will be more than happy to assist you to find your way.

In the book, I have tried to answer the questions that have continually come to my own mind during my travels in Ethiopia and those which have been put to me by fellow travellers. Why is Ethiopia so mountainous? Why are there so many deep gorges? What caused the Rift Valley to form and why does it have so many volcanoes? What causes the colourful hotsprings at Dallol and the spectacular lava lake of Erta Ale? How were people able to carve out the wonderful rock churches of Lalibela and produce the great obelisks of Axum? Why is Afar such an extraordinary and inhospitable desert of lava and salt? There are hundreds of such questions, and many of them have no definite answer. In attempting to provide answers, and at the same time produce a coherent and exciting story, I have tried to base my narrative on the most recent and/or most widely accepted thinking, without confusing the reader with technical details or by giving equal space to alternative interpretations. This has not been easy, and it is certain that many of my statements will be challenged by those who have studied the topics in detail and made different interpretations. My defence is that my purpose is not to provide a rigorous scientific text but to paint a broad picture which will engage rather than confuse the reader. A list of references is given at the end of the book, which the interested and determined reader may consult for further detail in order to form his or her own opinion.

The book is arranged in three parts. In order to make sense of what follows, the reader will find it helpful to have an overall picture of how the earth works and how geological time is measured. The first part, in Chaps. 1–3, aims to provide this as briefly and simply as possible, and to introduce some of the rock types that we will meet in the following pages. It concludes in Chap. 4 with a summary of Ethiopia's present-day physiography and geology. The second part, in Chaps. 5–8, outlines the geological story of Ethiopia, since her oldest rocks formed more than half a billion years ago to the present time. The third part, which comprises the bulk of the book in Chaps. 9–23, brings this story alive by travelling through the country and observing how it has shaped her rocks, scenery and even the history of her people. Chapter 24 gives a brief insight into the significance of earthquakes in Ethiopia, Chap. 25 attempts to bring everything together, and Chap. 26 speculates upon what may happen in the future. Maps, diagrams and photographs are provided to enable the reader, whether in the country or not, to locate himself or herself both physically and geologically. For those readers who are not in the country, I hope that these may help you to feel that you are!

A Note About Ethiopian Place Names

There is no standard system for the transliteration of Ethiopian place names. Even the name of the capital city is spelled in at least four different ways (Addis Ababa, Addis Abeba, Adis Abeba, Adis Abbaba), depending on which map one looks at.[1] Throughout the book, I have used the spelling which is most familiar to me, or which most closely resembles the Amharic or local ethnic phonetic. The names should be easily recognisable to those who would spell them differently. A more challenging complication is that many places have more than one name—an Amharic one and an ethnic one—and each of these can sometimes vary depending upon who your informant is! Current movement is towards the replacement of Amharic names in non-Amhara regions by their traditional ethnic ones. This is particularly so in the case of the Oromia region and can cause much confusion, particularly as many maps have not been updated. Again, I have used the name which is more familiar to me, and where appropriate given the alternative name in brackets.

The names of geographical features such as lakes, mountains and rivers can be even more problematic. Asking passers-by for the name of such a feature can often result in a variety of answers. Sometimes, the same feature is named differently by different groups of people or may not have a name at all. It is just a "river" or a "lake". An example of this is Lake Hayk in the Ethiopian highlands north of Dessie—"hayk" is simply the Amharic word for "lake". I have tried to name such features as correctly as possible and apologise for any misunderstandings.

There is even some confusion over the name of the country itself. Even today, many people still refer to her as Abyssinia, and historical texts frequently use the two names interchangeably. In fact, the name of Ethiopia for the country was not formally recognised internationally until after the establishment of the United Nations in 1945. The Ethiopia of the Bible is a Greek translation of the original Hebrew name of Cush. This is particularly misleading, since the land of Cush did not correspond geographically to present-day Ethiopia.

Finally, throughout the book, I have followed the practice of referring to Ethiopia in the female gender, as Ethiopians do in recognition of the fact that she is their motherland.

---

[1]The town of Wukro in northern Ethiopia holds the record for different transliterations of its name—with fourteen versions!

Part I
Setting the Scene

Since I want this book to be enjoyed by readers who have no geological background, I have tried to condense a few essential concepts into three short chapters. Chapter 1 explains the basics of how the earth works and where Ethiopia fits into the big scheme of things. Chapter 2 introduces the concept of geological time and the terms which are used to describe it, and Chap. 3 introduces the main rock types that we will meet during our journey around Ethiopia.

In Chap. 4, we will come to Ethiopia herself and take an overall look at her present topographic regions and how these relate to her geological make-up.

The Big Picture

1

In order to understand Ethiopia's geology and landscapes we first need to see where she fits in the overall scheme of how the earth works for, as will become apparent, she occupies a unique position. Inevitably, some terms will be introduced here that may not be familiar to you, but which will be very useful when you come to read the chapters which follow. Such terms are indicated in bold, and I have tried to explain them as clearly as possible.

1.1 The Earth's Interior

To begin with, we need to take a look inside the earth. It is made up of three main layers, illustrated in Fig. 1.1: the core, mantle and crust.

The **core**, shown in brown, is made of iron and nickel and is extremely dense. It has a radius of about 3400 km, more than half the radius of the whole earth. Its inner part is solid, surrounded by a liquid outer layer. The core will concern us very little in this book.

The **mantle**, shown in shades of orange and yellow, is formed of rock material which is rich in heavy metals such as iron and magnesium, and which is therefore denser than most of the rocks found at the earth's surface. It is about 2900 km thick and can be subdivided into three layers: the **lower mantle**, the **asthenosphere** and the **uppermost mantle**. Little is known about the lower mantle since it is so deep within the earth, but it is believed to be quite rigid since, although hot, it is under a

© Springer International Publishing Switzerland 2016
F.M. Williams, *Understanding Ethiopia*, GeoGuide,
DOI 10.1007/978-3-319-02180-5_1

Fig. 1.1 The main divisions of the earth's interior

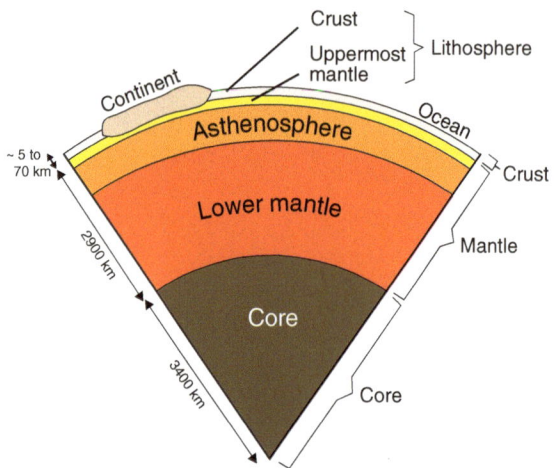

great deal of pressure. The **asthenosphere**, though also solid, is able to flow a little, rather like soft plastic, and contains pockets of molten material. Both these properties play an important role in what is happening at the earth's surface. The **uppermost mantle** is formed mainly of dense, rigid rock.

The **crust**, shown in grey and light brown, is the outermost layer of the earth and is composed of the rocks with which we are familiar: granite, basalt, sandstone, limestone and so forth. There are two kinds of crust. **Continental crust** (shown in light brown) forms the continents and is thick (on average around 35 km) and relatively light. **Oceanic crust** (shown in grey) lies beneath the oceans and is much thinner (around 6 km) and denser than the continental crust, though less dense than the uppermost mantle. In comparison to the earth as a whole, the crust is extremely thin. If we compare the earth to an egg, the crust would be on average only about one fifth the thickness of its shell.

The crust sits upon the uppermost mantle and is quite tightly connected to it. In many geological processes these two layers act as a unit and together are called the **lithosphere**. "Lithosphere" is in fact a more useful concept than the more familiar one of "crust" when discussing the large-scale movements of the earth, and will appear quite frequently in this book. Please try to get used to it—just remember that it means "crust plus uppermost mantle" and is essentially the earth's outer solid skin.

1.2 Moving Plates

The lithosphere probably solidified quite soon after the formation of the earth itself, about 4600 million years ago ("soon" in geological terms being just a few hundred million years!). Ever since then it has been in a continual state of motion. An analogy could be a saucepan of thick soup or porridge being heated very slowly on a stove. A scum or crust may form on the top which, as the soup or porridge simmers, breaks apart and moves around the surface. The earth's lithosphere acts in much the same way, albeit on a far vaster scale, and the process by which it does so is known as **plate tectonics**. This concept provides a very neat framework for understanding many things about the way the earth works, and particularly for understanding Ethiopia. If you don't grasp all the details right away don't worry—it took earth scientists over 40 years even to accept it as a serious proposition, and they are still gradually piecing it together.

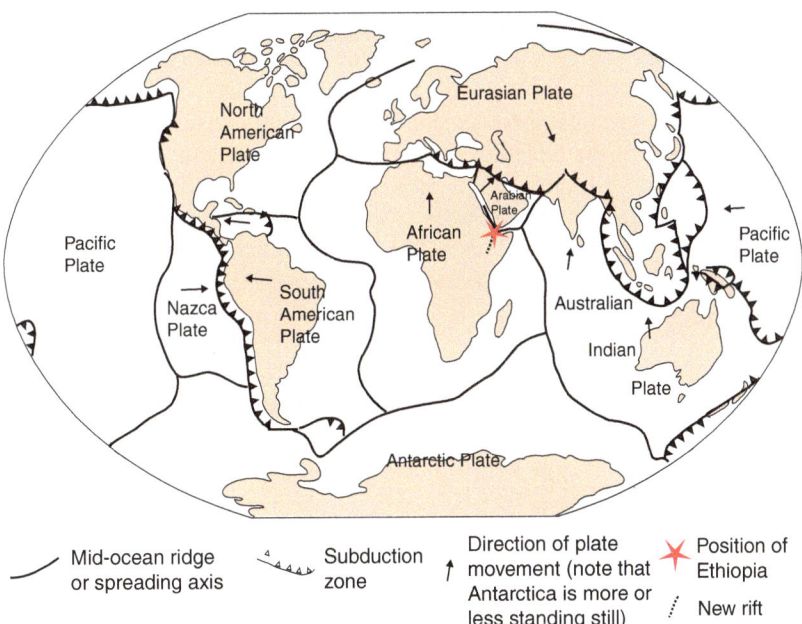

Fig. 1.2 The earth's tectonic plates as they are at the present time. Ethiopia's position is indicated by a *red star*

The basic premise of **plate tectonics** is that the lithosphere is not one continuous layer but is broken into fragments called **plates**. It is rather like a cracked eggshell, except that pieces of eggshell do not move about whereas the earth's plates do, so that their arrangement is continually changing. Their current arrangement is shown in Fig. 1.2. Some plates, such as the Pacific Plate, consist of oceanic lithosphere (uppermost mantle topped by oceanic crust) only; others, such as the African Plate, consist of both continental and oceanic lithosphere.

1.3   Plate Boundaries

The plates are separated from one another by **plate boundaries**. There are several types of plate boundary depending on how the plates are moving. These are illustrated in Fig. 1.3. Some plates, for example the South American and African ones, are moving away from each other (Fig. 1.3(i)). The widening gap between them becomes filled by molten material welling up from the asthenosphere to create new ocean floor. This material rises along a line of vents and fissures, forming a volcanic ridge known as a **mid-ocean ridge** or, more generally, a **spreading axis**, and the whole process is known as **sea-floor spreading**.

Where two plates are moving toward each other one of three things may happen. If oceanic lithosphere is meeting continental lithosphere, as in the case of the Nazca and South American Plates, the denser oceanic lithosphere will slide beneath the lighter continental lithosphere and be absorbed back into the mantle (Fig. 1.3(ii)). This process is known as **subduction**. As it descends, the slab of lithosphere becomes hot and partly melts, and the molten material rises to the surface as a volcanic eruption. If oceanic lithosphere meets oceanic lithosphere, for example the Pacific and Eurasian Plates along the western side of the Pacific Ocean, the situation is more complicated (Fig. 1.3(iii)). One side, generally that furthest from the continent, slides beneath the other, but at the same time gradually "rolls" backwards. Hot molten material rises to the surface as the descending lithosphere begins to melt, forming lines of volcanic islands off the coast. These are known as **island arcs**; examples are the Indonesian and Japanese islands. New ocean floor is created by molten material rising from the asthenosphere to fill the gap as the descending lithosphere rolls backwards.

When continental lithosphere meets continental lithosphere, such as the Australian-Indian and Eurasian Plates, the result is dramatic! Since both are equally light one side does not easily slide beneath the other. Instead they squeeze, buckle and compress each other—rather like a bad, but very slow, collision between two

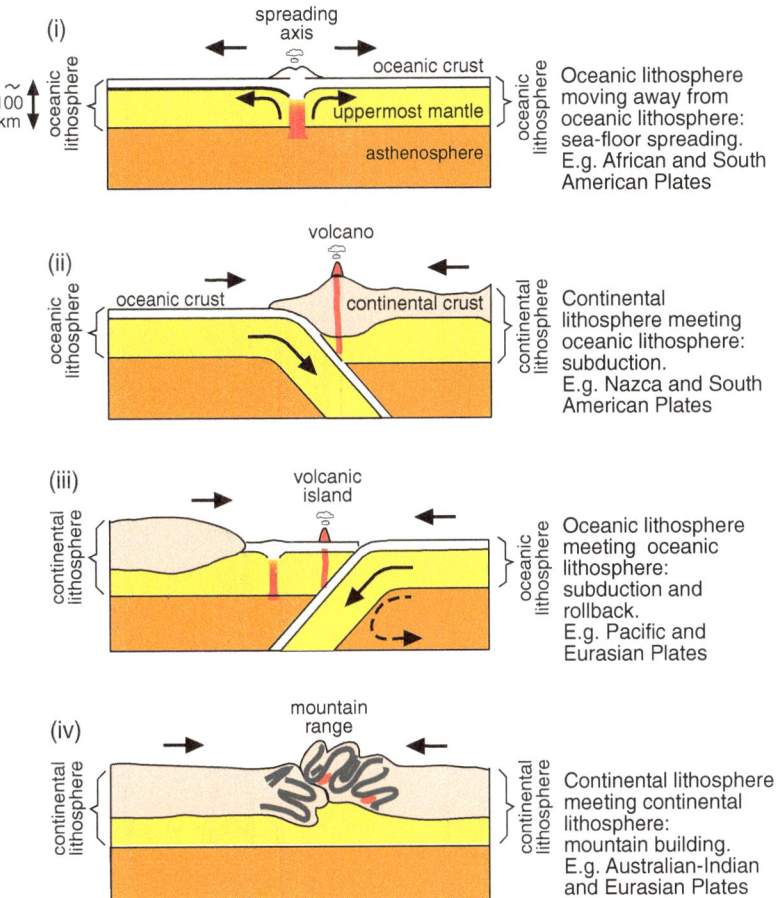

Fig. 1.3 The different types of plate boundary

cars. Eventually one side partially slides beneath the other, the whole process resulting in a massive mountain range of folded, fractured and uplifted rock—in this example the Himalayas (Fig. 1.3(iv)).

Another type of plate boundary, not shown in Fig. 1.3, is one where two plates slide past each other. The best known example of this is the San Andreas Fault in California, where the Pacific Plate is sliding alongside the North American Plate.

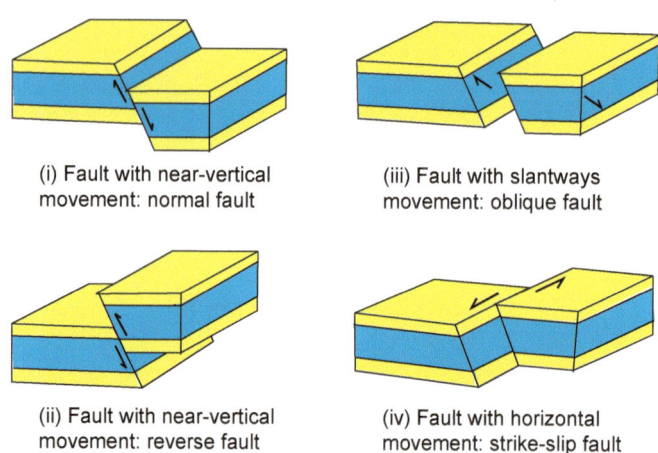

(i) Fault with near-vertical
movement: normal fault

(iii) Fault with slantways
movement: oblique fault

(ii) Fault with near-vertical
movement: reverse fault

(iv) Fault with horizontal
movement: strike-slip fault

Fig. 1.4 Types of fault

It is important to remember that all these processes take place extremely slowly,
a point which we will return to in the next chapter when we look at geological time.

A big question is: what makes the plates move? Various theories have been put
forward to answer this but as yet there is no consensus. One theory is that the
process more or less drives itself. Plates are pulled by the weight of the slabs of
lithosphere which are being subducted, and pushed by the new ocean floor being
created between them. Like many questions in geology, however, it remains wide
open to debate.

Two further terms, "**fault**" and "**tectonic**", will occur throughout this book, and
require a brief explanation. A **fault** simply means a fracture in the earth's crust, on
either side of which rocks have moved. The movement may be vertical or
near-vertical, horizontal or slantwise, as shown in Fig. 1.4. The term **tectonic** refers
broadly to processes and structures that result from movements within the earth.

1.4 Where Is Ethiopia in All This?

The red star on Fig. 1.2 shows where Ethiopia is located in the plate tectonic
scheme. You can see that she is situated just where the Arabian Plate has begun to
move away from the African Plate. That in itself is exciting enough—but even

more exciting is that Ethiopia herself is beginning to break apart, along the black dotted line on the map. This is the only place in the world where continental lithosphere is splitting away from continental lithosphere, and where the beginning stage of sea-floor spreading can be seen—on land! That is why Ethiopia is such a thrilling place for a geologist.

Geological Time

2

Geological time is incomprehensibly long. When a geologist speaks of something happening over a short period of time, he probably means that it took a million years or so to happen. Compared to the age of the earth, which is just over four and a half billion years, this is indeed a short period but in a human timeframe it is vast. It is easy for geologists to become complacent about the immensity of geological time, and difficult for non-geologists to assimilate it. An analogy that may help is to picture the history of the earth as a calendar year, in which the earth formed at midnight on the 1st January. Multicellular life appeared during mid-November and mammals on the 15th December. The dinosaurs died out on the 27th December and *Homo sapiens* (us!) appeared at 11.40 pm on the 31st December. On this scale the life of an average person would be about half a second!

Also, geological processes are overall very slow. Geological texts (and this book) are full of statements such as "the continents collided", "Australia split from Antarctica", "a mountain range was uplifted", "the sea advanced across the land". Processes like these actually take place over many thousands, millions, or even hundreds of millions of years. Australians, had they been around at the time, would not have noticed Antarctica disappearing into the distance. Continents do not collide with a single great crash. For example India is currently colliding with Asia, forming the Himalayan mountain range—and has been doing so for the past 50 million years. Rapid geological events such as earthquakes, volcanic eruptions or landslides certainly occur, but are merely jolts in the overall slow processes.

Because of its vastness, it is impossible to compress a true feeling for geological time into the time it takes to read a book. The narrative must inevitably be rushed.

© Springer International Publishing Switzerland 2016
F.M. Williams, *Understanding Ethiopia*, GeoGuide,
DOI 10.1007/978-3-319-02180-5_2

As you read, it will sometimes seem that the continents are scurrying over the earth's surface like ships ploughing across the ocean and bumping into each other, or that the sea is sweeping over the land like a tidal wave. From time to time it is a good idea to pull back and remember that all this is very, very slow and gradual. A good plan is to look at your little fingernail—it is growing at about the same rate at which continents move, a mountain range is uplifted or sea invades a continent.

2.1 The Geological Time Scale

Because geological time is difficult to measure precisely, and until the first part of the 20th century could not be measured numerically at all, geologists often prefer to speak in terms of geological eras, periods and epochs rather than numbers. It is similar to the way that we may speak of something as happening during the Middle Ages, or during the Axumite period, rather than stating its precise date. Over many decades a geological time scale has been constructed along these lines, and it is worth taking a little time to examine it since the terms it uses will inevitably crop up throughout this book.

The geological time scale, illustrated in Fig. 2.1, is built largely on the basis of life and evolution. The first major boundary is defined by what was believed at the time to be the first appearance of life on earth. The span of time before life appeared was termed the Precambrian era and that following its appearance was divided into three eras: the Palaeozoic (ancient life), Mesozoic (middle life) and Cenozoic (new, or recent, life). Although it is now known that simple life forms did exist during the Precambrian era, the boundary between it and the Palaeozoic certainly represents the sudden (in geological terms!) appearance of an amazing variety of complex life. An earth almost devoid of life became one teeming with it.

The Palaeozoic, Mesozoic and Cenozoic eras are divided into periods. Most of the period names, such as Cambrian, Ordovician and so forth, are rooted in early geological studies undertaken in Europe. The boundaries between the periods are defined by the appearance or disappearance of particular life forms in the fossil record. For example, the boundary between the Silurian and Devonian periods is defined by the appearance of a certain species of graptolite—a microscopic creature that lived in colonies resembling leaf impressions. Some boundaries are marked by the disappearance of a very large number of species. This is referred to as a mass extinction. The best-known mass extinction is that marking the boundary between the Cretaceous and Tertiary periods, when the dinosaurs and their

Fig. 2.1 The geological time scale (current in 2015). Eras are shown in *green*, periods in *yellow* and epochs in *orange*. Note that it is not drawn to scale. If it were, the Precambrian section would not fit on the page as it would be almost a metre long

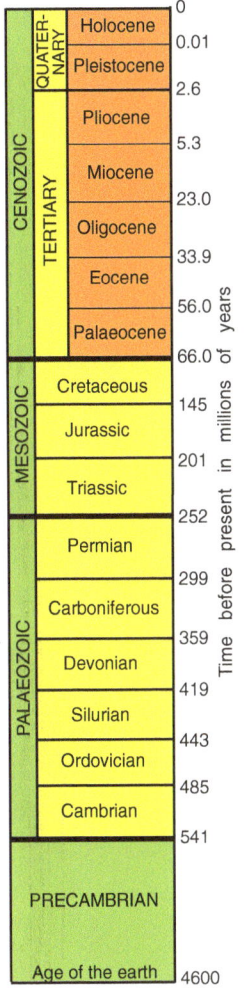

relatives, and a great number of other species living at that time, disappeared. The Quaternary period is not defined by life forms, but by the onset of the great ice ages when glaciers and ice sheets advanced and retreated over much of Europe and North America.

The Tertiary period[1] is further subdivided into five epochs whose names reflect the type or number of new species that emerged. Their names all end in "cene", from a Greek word *kainos* meaning "new". Thus Palaeocene means ancient new (species); Eocene, the dawn of new (species); Oligocene, few new (species); Miocene, less new (species); Pliocene, more new (species). The Quaternary period is subdivided into two epochs on the same basis: the Pleistocene, most new (species), and Holocene, wholly new (species).

It was not until the mid-20th century that actual ages, in numbers of years, came to be assigned to the eras, periods and epochs. This was thanks to the discovery of radioactivity, and the realisation that tiny amounts of radioactive elements contained in many types of rock acted as a kind of geological clock. Basically, the more a radioactive element has decayed, the older is the rock which contains it. As measurement techniques became more sophisticated, ages became more precisely determined. Even today the ages defining the boundaries on the geological time scale are continually being revised, updated and more significant figures added. The geological time scale illustrated is the current one at the time of writing.

---

[1]Shortly before this book was written the International Commission on Stratigraphy formally abolished the term "Tertiary", replacing it by the Palaeogene (Palaeocene, Eocene, and Oligocene) and Neogene (Miocene and Pliocene). In my opinion this has served merely to confuse something which is already sufficiently complicated and I will continue here to use the old and more familiar term Tertiary.

Some Notes About Rock Names

There are literally hundreds of names for different types of rocks. This chapter introduces just those few that you will come across while reading this book.

To begin with, rocks are classed into three main groups: sedimentary, igneous and metamorphic. **Sedimentary rocks** are formed of material that has been deposited by some kind of physical medium such as water, ice or wind, or by chemical precipitation, and then hardened by becoming cemented or compressed. The main sedimentary rocks that we will come across in this book are sandstone, which is formed of cemented grains of sand, and limestone, which is formed of the chemical calcium carbonate.

Igneous rocks are those which have formed by solidification of molten material. Molten rock material in general is called magma, but when it reaches the surface of the earth, for example during a volcanic eruption, its name changes to lava. This is a little confusing, made more so by the fact that the rock formed when the lava solidifies is frequently also referred to as lava.

Igneous rocks are broadly classified according to three criteria: the quantity of silica (silicon dioxide) they contain, the type of feldspar mineral present, and their grain size. These criteria may seem a little puzzling, but there is good reason for them. Oxygen and silicon are by far the most abundant elements in the earth's crust. The mineral quartz is formed entirely of oxygen and silicon, and almost all the other common rock-forming minerals are composed of these two elements in combination with aluminium, iron, calcium, sodium, potassium or magnesium.

Feldspars comprise the most abundant group of igneous rock-forming minerals. They are compounds of silicon, oxygen, aluminium and one or more of the elements calcium, sodium and potassium. They fall into two main categories:

© Springer International Publishing Switzerland 2016
F.M. Williams, *Understanding Ethiopia*, GeoGuide,
DOI 10.1007/978-3-319-02180-5_3

potassium feldspars which contain, obviously, potassium, and plagioclase feldspars which contain calcium, sodium or both.

The type of feldspar and the quantity of silica provide information about the origin of the magma from which the igneous rock formed, and the grain size indicates how quickly it solidified. Coarse grains indicate that the rock solidified slowly (giving the grains more time to grow), generally beneath the earth's surface. Such rocks are termed intrusive. The most familiar example of an intrusive igneous rock is granite. Fine grains tell that a rock solidified quickly, which happens when magma reaches the earth's surface (and becomes lava). Such rocks are termed extrusive, or volcanic since they frequently erupt from volcanoes. The most familiar example of an extrusive or volcanic rock is basalt. Sometimes lava

Table 3.1 Summary of the main igneous rocks encountered in this book

Amount of quartz		No quartz and usually high in magnesium and iron	Very little or no quartz	At least 20% quartz
Mafic or silicic		Mafic (high **Ma**gnesium and **Fe**rrum (iron))	Intermediate	Silicic
Predominant type of feldspar		Calcium feldspar	Potassium feldspar	Potassium feldspar
Texture	Coarse grained	GABBRO	SYENITE	GRANITE
	Fine grained	BASALT	TRACHYTE	RHYOLITE
	Glassy			OBSIDIAN
	Partly glassy, some crystals			PITCHSTONE
	Fragmented — Loosely packed fragments	TUFF	TUFF	TUFF
	Fragmented — Compressed fragments			IGNIMBRITE
	Fragmented — Welded fragments			WELDED TUFF
	Many gas bubbles (vesicles)	SCORIA		PUMICE

solidifies so quickly that no grains at all have time to grow and the rock is completely glassy; other times it traps many gas bubbles as it solidifies, and is frothy. The most familiar example of a frothy igneous rock is pumice. Table 3.1 summarises the main igneous rocks that we will encounter in this book.

Metamorphic rocks are those whose composition, texture or both have been altered (*meta* = change; *morph* = form) by heat and pressure, generally due to earth movements or burial deep within the earth. Their composition and texture (and hence name) depend on what the original rock was before it was metamorphosed, and the degrees of temperature and pressure which caused the metamorphism. We will encounter metamorphic rocks mainly in Chaps. 10 and 18, and specific names will be introduced as they occur. It is handy to note, however, that metamorphic rocks are often referred to simply by prefixing "meta" to the rock type that they were before they became metamorphosed; for example metasediment (for a metamorphosed sedimentary rock), metavolcanic (for a metamorphosed volcanic rock), metabasalt (for a metamorphosed basalt) and so forth.

Ethiopia Today

4

Before exploring Ethiopia's geological past it is useful to take a look at her present topographic regions and see how these relate to her geological make-up. In doing this, and in many of the following sections of this book, we come up against the inconvenience of politics. Geology has no respect for national boundaries and Ethiopia and her neighbours, particularly Eritrea and Djibouti, are a geological and geographical entity and will be considered as such in the discussions which follow. Since it would be cumbersome to use a term such as Ethiopia/Eritrea/Djibouti throughout the book, the reader should be aware that the term Ethiopia often implicitly comprises the region covered by the three countries. The currently fragmented lands which used to constitute Somalia are also relevant to parts of our story, but logistical reasons prevent their being included in any detail.

4.1 Ethiopia's Topographic Regions

Ethiopia can be divided roughly into the five topographic regions shown in Fig. 4.1: the Western Highlands, the Southeastern Highlands, the Ethiopian Rift Valley, Afar and the Ogaden, all with very different scenery, climates, and even people. The highlands are often referred to as plateaux, but this term is misleading as it implies something that is flat on top, like a tableland. Although substantial stretches of the Western Highlands are plateau-like, equally large areas are far from being so, and the Southeastern Highlands form a narrow ridge rather than a plateau. A British soldier who accompanied the Napier expedition to Ethiopia's Western Highlands in 1868 wrote that "if this is a tableland they have turned the table

© Springer International Publishing Switzerland 2016
F.M. Williams, *Understanding Ethiopia*, GeoGuide,
DOI 10.1007/978-3-319-02180-5_4

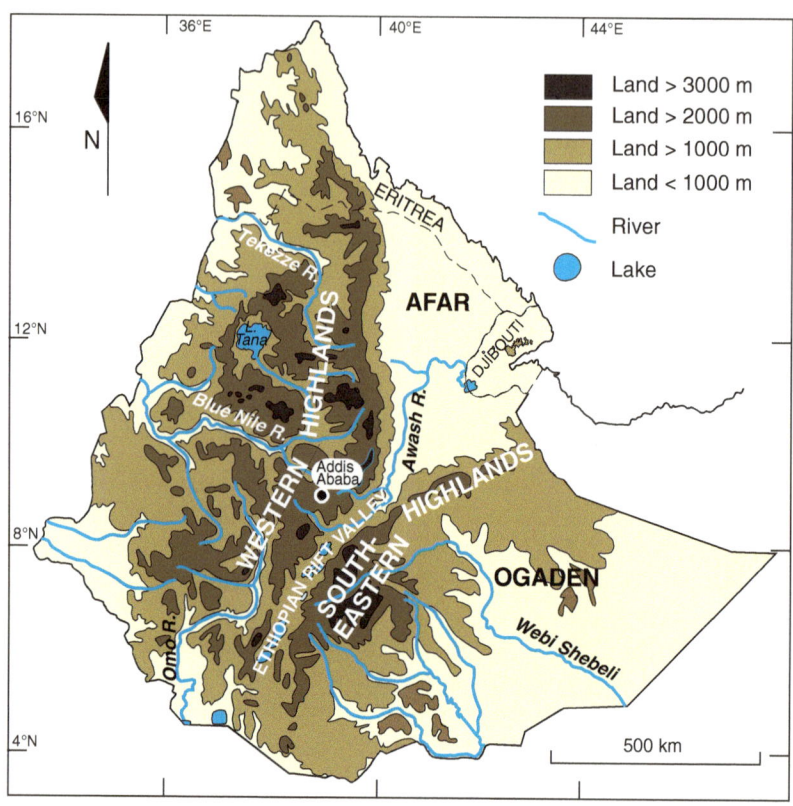

Fig. 4.1 Ethiopia's main topographic regions

upside down and we are scrambling up and down the legs". This is an apt observation. The highlands range from rugged and majestic peaks reaching elevations of over 4500 m, to gently rolling fertile hills and plains which are dissected by deep and often impassable gorges. The Western Highlands taper away westwards toward the Sudan plains, and the Southeastern Highlands southeastwards to the Ogaden, but to the east and north respectively they terminate abruptly at high escarpments which separate them from Afar.

The Western and Southeastern Highlands are separated from each other by the Ethiopian Rift Valley (also known as the Main Ethiopian Rift), a strip of land some 80 km wide and 1500 m lower than the highlands on either side, and generally bordered by steep escarpments. At its northern end the Rift Valley opens out into

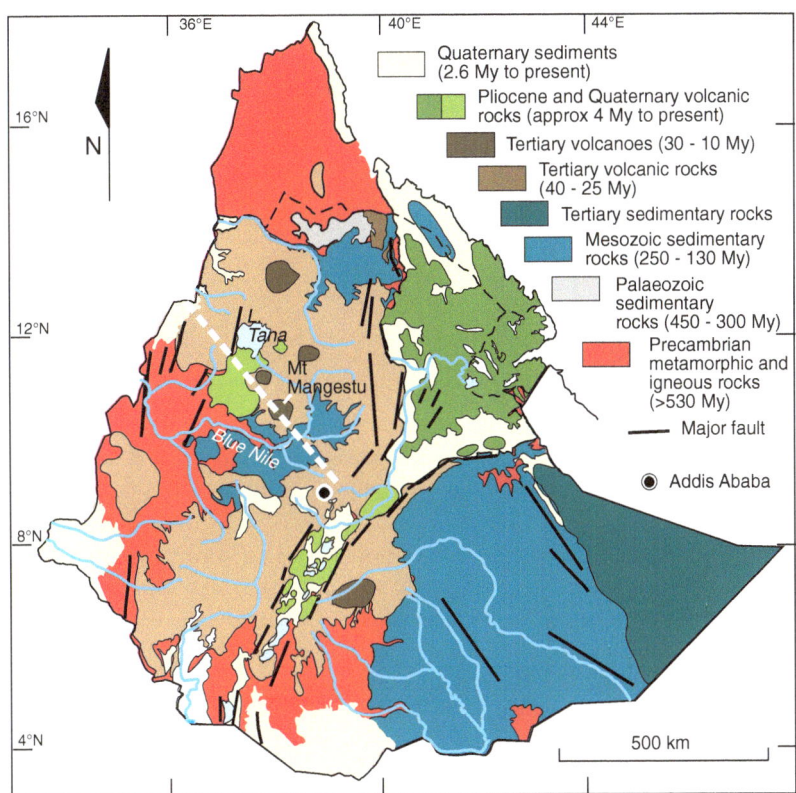

Fig. 4.2 Simplified geological map of Ethiopia. The abbreviation My in the legend stands for millions of years before the present time. Modified from Geological map of Ethiopia 1:2,000,000 2nd ed. (1996), Geological Survey of Ethiopia

Afar, a low-lying desert region, parts of which lie below sea level. The Ogaden is a region of flat, dry plains extending into Somalia.

4.2 The Geological Map

The geological map in Fig. 4.2 indicates how Ethiopia's geology relates to these topographic regions. Even if you are not accustomed to looking at geological maps, you can see how the colours on it correspond more or less to the regions

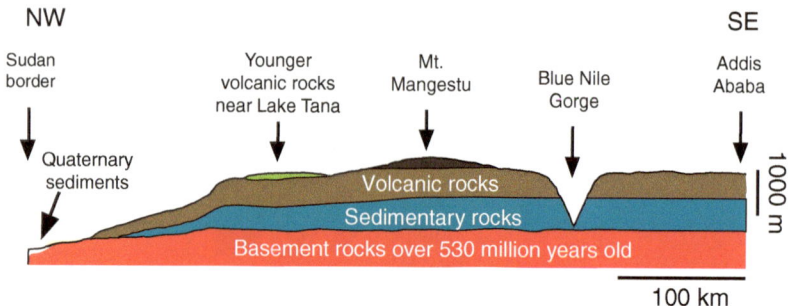

Fig. 4.3 A schematic geological cross section through the Western Highlands, along the *dashed line* shown in Fig. 4.2. Note that the vertical scale is greatly exaggerated

shown on the topographic map (Don't worry about the individual rock formations named on the legend; you will become familiar with these as we proceed through the book).

The brown areas represent layers of volcanic rock, the result of a period of great volcanic activity which occurred around 30 million years ago when flows of lava (molten rock) poured out across large parts of the country. These volcanic rocks cover most of the highland regions, and are responsible for Ethiopia's fertile soils.

Blue and grey areas represent sedimentary rocks, mainly sandstones and limestones. These were laid down by rivers and glaciers, and beneath a sea which covered Ethiopia long before the eruptions which produced the volcanic rocks. Over much of the country the sedimentary rocks are buried beneath the volcanic ones. They are exposed mainly where rivers such as the Blue Nile have cut deep gorges into them; and in the Ogaden region where the cover of volcanic rocks was thin and, apart from patchy remnants, has been eroded away.

Red represents Ethiopia's oldest rocks, formed more than 530 million years ago. Because these lie underneath everything else, they are rarely exposed other than in the deepest river gorges, and around the edges of the country where the overlying rocks have been worn away and in some places may never have been deposited. These oldest rocks are often referred to as the basement, because all the other rock formations are built on top of them.

Green areas represent Ethiopia's youngest rocks which result from volcanic activity between 4 million years ago and the present time. These occur mainly in the Rift Valley and Afar, and include many volcanoes which are either dormant (meaning that they might erupt) or active (meaning that they are actually erupting).

Finally, pale yellow represents patchy coverings of loose material such as sand, deposited mostly in lowland regions by wind and water during recent times.

Of course all the rocks shown on the geological map cannot actually be seen at the earth's surface. With the exception of the Afar region, they are for the most part covered by soil and vegetation so that the map has to be pieced together from the rocks that can be seen, called outcrops. It is like the children's puzzle of "joining the dots" to find out what the picture shows. A geological map takes many years of careful geological work to construct, and even then is continually being updated as geological exploration continues.

Even when the map has been constructed, it has the shortcoming of showing only the uppermost layer of rocks. It does not show what lies below them. We can picture it as a layer cake in which we can see only the icing on the top. In this analogy, Ethiopia would be a rather messy layer cake. In some places the icing (the layers of volcanic rock) did not reach all the way to the edges of the cake, and even there has been nibbled away quite a bit. In the Ogaden it has been pretty much licked off altogether. The cake has sagged and cracked across the middle (the Rift Valley), and one big piece has been bitten right off (Afar). But fortunately someone has cut a couple of quite neat slices which reveal the layers underneath the icing (the deep gorges of rivers such as the Blue Nile and the Tekeze). By piecing together the information revealed by the slices and the uncovered bits around the edges, we could work out a picture of what the interior of the cake would look like. Similarly, geologists can use what is exposed at the earth's surface to work out how things are arranged below, and thus construct a cross section which gives a vertical view of the geology. Figure 4.3 shows such a cross section, through the Western Highlands from Addis Ababa to the Sudan border, along the dashed line shown in Fig. 4.2.

In the next four chapters we will see how these layers of rock were formed, and how Ethiopia was sculpted by geological processes into the country we see today.

Part II

Ethiopia's Geological Story

The next four chapters outline Ethiopia's geological story, from when her oldest rocks formed more than half a billion years ago to the present time. As you read, it is important to be aware that many parts of the story are still being worked out, and there may be gaps, contradictions, loose ends and open questions. Of course, this can be frustrating, but on the other hand, it is what makes geology so exciting. It is like trying to put together a jigsaw puzzle from which many pieces are missing, but in which the final picture becomes gradually clearer as more and more pieces are found.

The problem is that much of geology is concerned with events that took place in the deep past when no one was around to witness them and with features which are far beneath the earth's surface where no one can see them. A great deal of geological understanding must therefore be obtained by inference, using clues from what can be seen and what can be measured. Fortunately, there are many such clues—not only the rocks and landforms which can be observed and studied at the earth's surface, but also those provided by indirect methods such as geophysics (e.g. studying earthquake waves which can penetrate deep beneath the earth's surface), geochemistry (analysis of the detailed composition of rocks which can give clues as to their origin) and geochronology (methods for finding the ages of rocks, usually from the radioactive elements they contain). New techniques are continually becoming available and old ones refined, so that over the past few decades, our understanding of geological processes and of what is hidden beneath the earth's surface has increased greatly. However, since so much of the evidence is indirect, there are often different ways of interpreting it, and it frequently happens that even the most experienced researchers have different interpretations. This makes it difficult to tell a simple story of Ethiopia's geological history—firstly, because it is not simple but extremely complex and secondly, because there are many details upon which there is no consensus.

The following narrative will inevitably be oversimplified, but I hope that this may be outweighed by its being informative and clear.

Beginnings: The First Half-Billion Years

5

As ancient rocks go, the oldest rocks in Ethiopia are not very old. The oldest rocks found on earth are more than 4 billion years old, in the ancient continental cores of Australia, Canada and Greenland. Few rocks have been found in Ethiopia with ages greater than 850 million years. This is because, earlier than this, Ethiopia and almost all the rocks of which she is formed did not exist.

To understand this we need to go back a billion (1000 million) years. At that time all the earth's land surface was gathered in a single continent named Rodinia. Rodinia was almost certainly not the earth's earliest continent. Continental crust had formed and plate tectonics, explained in Chap. 1, had come into operation very soon (in geological terms!) after the formation of the earth 4600 million years ago. This early crust (or, more strictly speaking, lithosphere) had probably undergone several cycles of breaking apart and coming together before Rodinia was formed. But Rodinia, whose name comes from a Russian word *ródina*, meaning "motherland", is where our story will start.

It is not certain exactly how the components of Rodinia were arranged, but a possible scenario is shown in Fig. 5.1(i). About 900 million years ago Rodinia began to break apart. Cracks formed across the great continent (Fig. 5.1(ii)) and molten rock from the earth's mantle welled up through them, producing ocean floor between the broken fragments of continent as they moved apart (Fig. 5.1(iii)). The fragments, known as cratons, are still recognisable as the nuclei of today's continents and are labelled as such throughout Fig. 5.1. By about 850 million years ago all the cratons were separated by oceans. The one which concerns us here, the Mozambique Ocean, extended from the Congo to the Indian craton (Fig. 5.1(iv)).

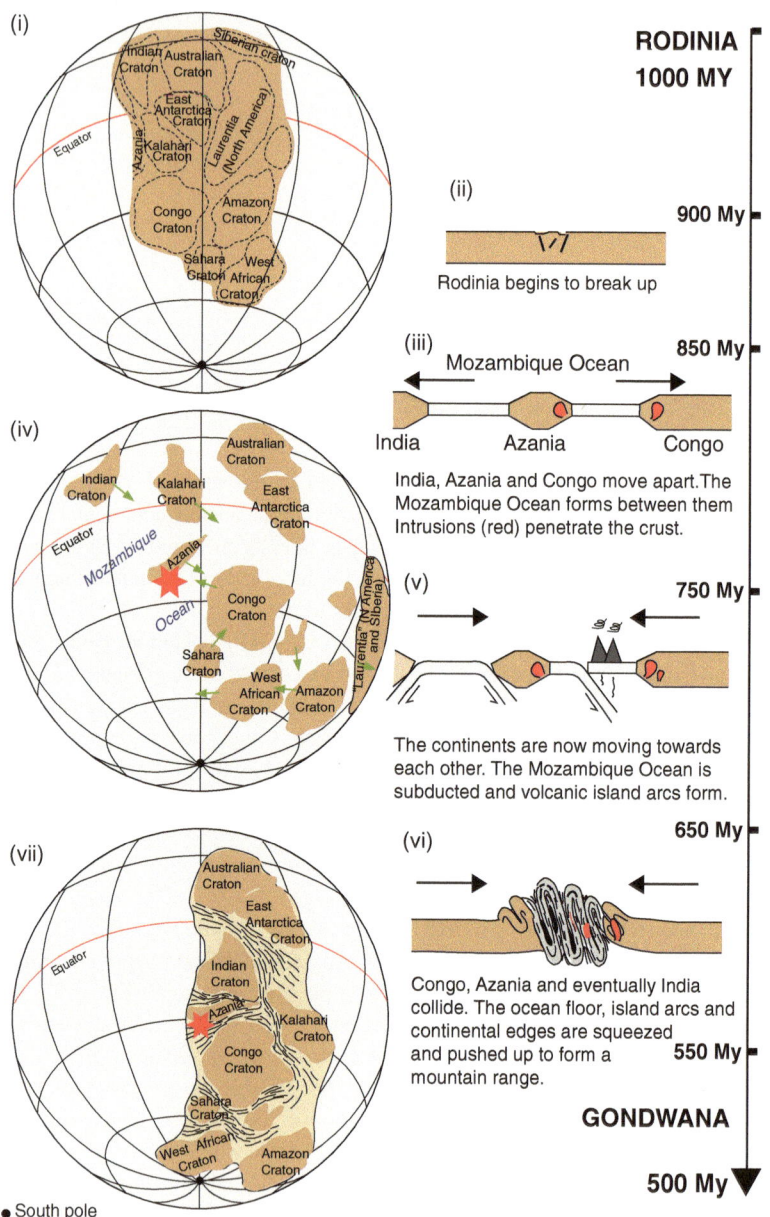

(i)

(ii)

Rodinia begins to break up

(iii)   **Mozambique Ocean**

India   Azania   Congo

India, Azania and Congo move apart. The
Mozambique Ocean forms between them
Intrusions (red) penetrate the crust.

(iv)

(v)

The continents are now moving towards
each other. The Mozambique Ocean is
subducted and volcanic island arcs form.

(vii)

(vi)

Congo, Azania and eventually India
collide. The ocean floor, island arcs and
continental edges are squeezed
and pushed up to form a
mountain range.

• South pole

RODINIA
1000 MY

900 My

850 My

750 My

650 My

550 My

GONDWANA

500 My ▼

◀ **Fig. 5.1** Time line illustrating Ethiopia's first half billion years, from the break-up of Rodinia to the formation of Gondwana. *Pale grey* represents oceanic crust, *light brown* continental crust and the *red blobs* are intrusions. *The red star* indicates the position of Ethiopia as Azania, India and the Congo craton moved toward each other and collided. The continental configuration diagrams are based on those of Collins and Pisarevsky (2005)

Around 700 million years ago all this changed. The cratons stopped moving apart and began moving back towards each other (Fig. 5.1(iv), (v)). The Indian craton and one called Azania, which corresponds roughly to today's Madagascar with bits of Somalia and Arabia attached, moved toward the Congo craton. The ocean crust between them was swallowed, or subducted, beneath the cratons, and volcanic island arcs formed as shown in Fig. 5.1(v). As the ocean became narrower the volcanic islands and the remaining ocean floor were squeezed together until, about 630 million years ago, Azania and then India finally came up against the Congo craton. The squashed ocean floor and volcanic islands, and finally the edges of the colliding cratons themselves, were squeezed, crushed and pushed upward to form a great mountain range (Fig. 5.1(vi)).

The crushing, squeezing and uplifting involved in this great collision was accompanied throughout by intrusions of molten rock (magma). In daily life an intrusion occurs when someone pushes in (generally where they are not wanted!), and its geological meaning is similar. Masses of molten rock generated by the heat and pressure of the mountain building event pushed their way upward into rock that was already there, and solidified before reaching the surface. Since they are produced by solidification of magma, intrusions are formed of igneous rocks and, since the magma cooled slowly, these are coarse-grained rocks such as granite.

The whole process of collision, intrusions and uplift was a long, slow one which lasted for over 100 million years. It has been termed the East African Orogeny, from Greek words *oros* meaning "mountain" and *genesis* meaning "creation". It was similar to the process which is forming the Himalayan mountain range today, where India is colliding with Eurasia, and the mountains formed would have rivalled the Himalayas in height and majesty. They would, like the Himalayas, have been built of a mix of metamorphosed igneous and sedimentary rocks, and igneous intrusions, bent into massive folds and sliced by faults where slabs of rock had fractured and been pushed over each other.

The mountain belt which resulted is known as the East African Orogen,[1] and this is where Ethiopia's oldest rocks were born.

---

[1]Note that "orogeny" is the process of mountain building; "orogen" is the resulting mountain belt.

By about 540 million years ago all the ancient cratons which form the cores of today's southern continents (South America, Africa, India, Antarctica and Australia) had come together by a similar process and were connected by mountain belts which marked their collision fronts. Together they formed a supercontinent known as Gondwana (Fig. 5.1(vii)). The cratons which constitute today's northern continents (North America, Europe, Asia) had similarly come together, forming a supercontinent called Laurasia. For a while Laurasia was joined to Gondwana forming a super-supercontinent known appropriately as Pangea (the whole earth).

The East African Orogen has two distinct sections with rather different rock types. The reason for this is that as the Congo craton, Azania and the Indian craton became closer together, the ocean crust between them was not only pushed upwards but was also squeezed out sideways towards the west where, as you can see in Fig. 5.1(vii), Gondwana had an open edge. It is rather like squeezing a tube of toothpaste—as the sides of the tube (the continental edges) come together the toothpaste (the ocean floor in between them) is pushed out of the nozzle. This meant that oceanic type rocks became concentrated at the westernmost end of the orogen. For a long time it was believed that these rocks were very old, like a craton, and they became known as the Arabian-Nubian Shield. They were not thought to be connected with the rest of the orogen, which was known as the Mozambique Belt. Even though it is now realised that they are both part of the same feature, the names have stuck. Figure 5.2 shows how the Arabian-Nubian Shield and the Mozambique Belt are related, and you can see that Ethiopia straddles them both.

The rocks formed during the East African Orogeny are very important to Ethiopia's economy, as most of her known mineral deposits are found within them. This is why the distinction between rocks of the Arabian Nubian Shield and those of the Mozambique Belt is important, for the former are by far the richer in economically valuable minerals. An exploration geologist must have a good understanding of where the different rock types occur, and how to recognise them, so that he or she saves time (and money!) by looking in the most likely places.

We can see now why almost all the rocks in Ethiopia are less than 850 million years old. The Mozambique Ocean and its volcanic island arcs, of which the Arabian-Nubian Shield part of Ethiopia is formed, did not exist before then, The rocks of the ancient continental edges, of which the Mozambique Belt part of Ethiopia is formed, were changed to new (metamorphic) ones by heat and pressure during the East African Orogeny. There are however just a few instances where

Fig. 5.2 The East African Orogen as it was in Gondwana, showing the Arabian Nubian Shield and the Mozambique belt. Ethiopia overlaps them both. At the time of its formation the Orogen would have been oriented E–W. The figure shows it rotated to its present N–S orientation

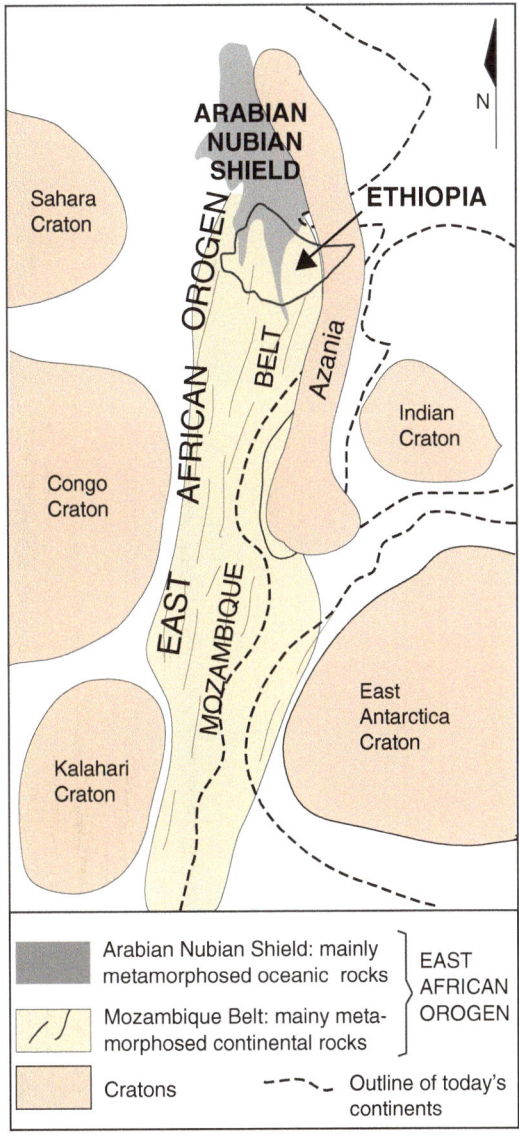

much older ages have been measured on Ethiopian rocks, even greater than 2000 million years. These ages were measured on minute crystals of a mineral called zircon, which is very robust and not as easily altered by heat and pressure as the common rock-forming minerals. Zircon crystals may continue to indicate the age at which they first crystallised from a magma, perhaps in one of the original cratons, rather than the age at which the rock containing them was metamorphosed. We shall see a little more of zircons in Chap. 10.

Although this great mountain building event occurred so long ago, it has continued to have a strong influence throughout Ethiopia's geological history, right up to the dramatic events which are happening today.

The Early and Middle Years: Ice, Sea and Sand

Following the upheavals of Ethiopia's first half billion years, things became geologically fairly quiet. At the start of the Cambrian period, about 540 million years ago, the southern continents were all joined as the supercontinent Gondwana as shown in Fig. 5.1(vii), and the land that was to become Ethiopia was part of the great mountain range of the East African Orogen. Gondwana remained intact for more than 300 million years. However, it did not remain still. Firstly it rotated as a whole by about 60° in a clockwise direction. This took about 40 million years—quite fast in geological terms! It also gradually drifted southwards. Figure 6.1 shows Gondwana as it may have been about 420 million years ago.

© Springer International Publishing Switzerland 2016
F.M. Williams, *Understanding Ethiopia*, GeoGuide,
DOI 10.1007/978-3-319-02180-5_6

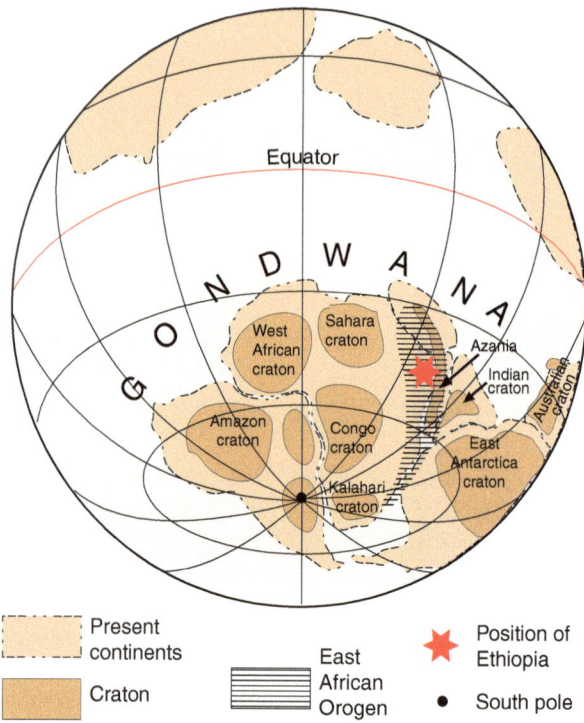

Fig. 6.1 Gondwana at about 420 million years ago. The *outlines* of the present continents are shown in *light brown*. The areas between the cratons are occupied by mountain ranges similar to the East African Orogen

6.1   The Palaeozoic Era: Blank Pages and Icy Spells

As Gondwana rotated and drifted, the mountains of the East African Orogen were slowly being worn down by erosion: the action of water, wind and, to a lesser extent, ice. The vast amounts of sediment washed down from the region which was to become Ethiopia were carried away by rivers across the future Sudan, northern India, the northeastern Arabian Peninsula, and out into the oceans beyond. The

PHOTO 6.1 Tillite from near the town of Wukro in northern Ethiopia. Tillite consists of rock fragments of different types and sizes, all mixed up in a fine, clayey matrix. All this material has been broken up, carried and dumped by ancient glaciers. The *pen* shows the scale (2011)

worn down remnants of the ancient mountains became the Precambrian[1] basement, which we saw shown in red on the geological map in Fig. 4.2—the base upon which the whole of the later geological sequence was to be built.

During most of the Palaeozoic era rocks were being worn down in Ethiopia and sediments were being carried away rather than being deposited. Very few new rock layers were laid down and there were long gaps for which no rocks exist at all. Rocks are what the geologist relies upon to record events, and if there are none it is difficult to work out what, if anything, was happening. Such gaps in the rock record are called unconformities. It is rather like trying to read a history book from which pages have been torn out, or were never written. We have to surmise what the missing pages might have told us.

Only two events are recorded during the entire Palaeozoic era, and they are rather unexpected ones. A type of rock known as tillite is found in several localities, mainly in northern Ethiopia, together with layers of sandstone. Tillite, illustrated in PHOTO 6.1, is composed of broken rock fragments in a matrix of fine clayey material and, by comparison with similar material forming at the present time in glaciated regions, is known to be characteristic of deposition by ice.

[1]This term is used rather loosely, as some of the basement rocks extend into the early Cambrian.

Its presence here indicates that there were two brief episodes, one during the late Ordovician period and the second during the Carboniferous and early Permian, when glaciers were active over parts of Ethiopia. This at first seems surprising because, particularly during the Carboniferous period, Gondwana had shifted so that Ethiopia would have been in what today are tropical latitudes. At that time however the whole global climate was much colder than it is at present, and during the Permian and Carboniferous periods a much larger ice cap than today's was centred on the South Pole. Glacial deposits of that age are found in many parts of Gondwana, some even further from the pole than Ethiopia was.

6.2   The Mesozoic Era: Sand, Sea and Early Warning Signs

A little before 250 million years ago Gondwana, which had been an intact and quite stable land mass for almost 300 million years, began showing signs of being ready to break apart. Cracks formed across some parts of the supercontinent, including those which were to become eastern and central Africa, and land between the cracks subsided to form rifted basins: broad, shallow, flat-floored valleys bordered by fractures, or faults (Fig. 6.2(i)). The probable locations of the rifted basins are shown in Fig. 6.2(ii). Their form and locations are somewhat schematic since in most places they are deeply buried beneath later sediments and volcanic rocks. They have been detected largely by geophysical methods, or in deep boreholes drilled for oil and mineral exploration. They are however important in the effect that they have upon later events, as we shall see.

Rivers flowed into these basins, carrying sediment eroded from the basement rocks and depositing it as sand along their valleys and across their deltas and flood plains. Over time (about 50 million years) the layers of sand grew to a thick pile and became hardened and cemented to sandstone. This is known as the Adigrat Sandstone. It is useful to note here that rock formations, particularly sedimentary rocks, are often named after the place where they were first noted and described, or where they can be particularly well seen. This sandstone takes its name from the town of Adigrat in northern Ethiopia, where it was first noted in 1867 by Thomas Blanford, the geologist who accompanied the British Napier Expedition on its mission to rescue British prisoners held by the then Emperor Tewodros. We shall encounter the Adigrat Sandstone a number of times in our journey around Ethiopia.

Although the cracks did not develop further than the stage of rifted basins, they were the earliest indication of the next major event in Ethiopia's geological story: the break-up of Gondwana and the break-up of Ethiopia herself. The supercontinent

Fig. 6.2 (**i**) Cross section through a rifted basin. (**ii**) Possible configuration of the rifted basins, shown by the stippled areas, which formed in the Ethiopia region during the late Palaeozoic and early Mesozoic eras. Note that the geographic setting is shown as it is at the present time. The Rift Valley, Afar, Red Sea and Gulf of Aden would not have been there when the rifted basins formed. Modified from Gani et al. (2008)

(i)

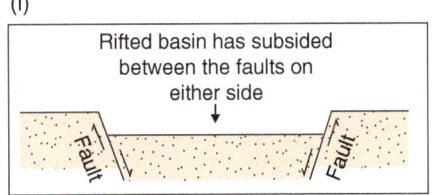

(ii)

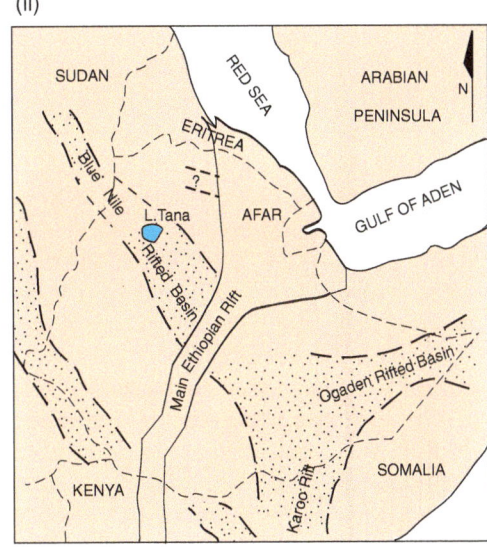

was beginning to split apart, just as Rodinia had done some 700 million years previously.

Around 180 million years ago, some 70 million years after these first cracks formed, the eastern part of Gondwana began to separate from Africa. India, Australia and Antarctica broke off and moved away eastward, and sea-floor spreading commenced in what was to become the Indian Ocean. The waters of this new ocean gradually extended northwestward across Ethiopia and her neighbouring lands, flooding and overflowing the rifted basins. At first it was a shallow sea which deposited mud, and layers of calcium carbonate and gypsum (hydrated calcium sulphate), minerals which form by evaporation of warm, shallow sea water. As the sea continued to move in and deepen, thick layers of limestone were deposited, some by chemical precipitation of carbonate minerals dissolved in the seawater and some from the shells of organisms living in the sea. The limestone is known as the Antalo Limestone, after the town of Antalo (now known as Hintalo) in northern Ethiopia.

The sea eventually extended westward to just beyond the present position of Lake Tana and northward almost to Eritrea, and covered much of the Arabian Peninsula. Towards the end of the Jurassic period, about 150 million years ago, it began to retreat, possibly because the land was beginning to rise as we will see in the next chapter. Dry land was once more exposed, and rivers flowed over it into the retreating sea, depositing layers of sand in much the same way that rivers had deposited the Adigrat Sandstone some 100 million years earlier. This later sandstone has been given various names according to different places where it has been found; for consistency I will refer to it as the Upper Sandstone.

The Ogaden region of southeastern Ethiopia is a little different, as it was covered by the sea for longer than the rest of Ethiopia. This is because, firstly, the sea was moving in from that direction so it was the first region to be submerged, and secondly, the land there was subsiding as the Indian Ocean widened. In fact the sea advanced over the Ogaden region, and withdrew again, a number of times after it had completely retreated from the rest of the country. The limestone layers are therefore much thicker in the Ogaden than they are anywhere else in Ethiopia.

The Mesozoic part of Ethiopia's history is thus recorded as a kind of sandwich, of sandstone—limestone—sandstone.

Toward the end of the Mesozoic era, about 70 million years ago, Ethiopia can be pictured as a land of mainly sandstone plains, probably without a great deal of topographic relief and not rising much above sea level. Sea still covered the Ogaden region but by the end of the Eocene epoch, about 35 million years ago, it had retreated from there also, almost to the present coastline of Somalia. There was no Red Sea or Gulf of Aden, and Ethiopia and her neighbours were still joined to the Arabian Peninsula. However, all this was about to change, as we will see in the next chapter.

The Onset of Turbulent Times

As we saw in the previous chapter, by the end of the Mesozoic era Ethiopia had undergone a number of changes following the great collision of continents which had brought her into being. The mountains of the East African Orogen had been eroded down to their roots, the glaciers of the Palaeozoic era had come and gone, the Mesozoic sea had advanced and retreated from everywhere except the Ogaden, and much of the old Precambrian basement was covered by layers of limestone and sandstone. For the first 30 million years or so of the Cenozoic era (during the Palaeocene epoch and much of the Eocene), Ethiopia and her neighbours remained much as they had been at the end of the Mesozoic: a land extending unbroken from beyond Ethiopia's western borders to the eastern side of the Arabian Peninsula.

The break-up of the rest of Gondwana, however, was well under way. The situation at about 60 million years ago is shown in Fig. 7.1. South America had broken away from Africa about 130 million years ago, and sea-floor spreading was creating the Atlantic Ocean between them. India had already split away from Africa as we saw in the previous chapter, and the Indian Ocean was widening as new sea floor was created along a spreading axis called the Carlsberg Ridge. Once-united Gondwana had split into four separate plates: the African Plate, the South American Plate, the Indian Plate and the Antarctica-Australian Plate (Australia had not yet separated from Antarctica).

© Springer International Publishing Switzerland 2016
F.M. Williams, *Understanding Ethiopia*, GeoGuide,
DOI 10.1007/978-3-319-02180-5_7

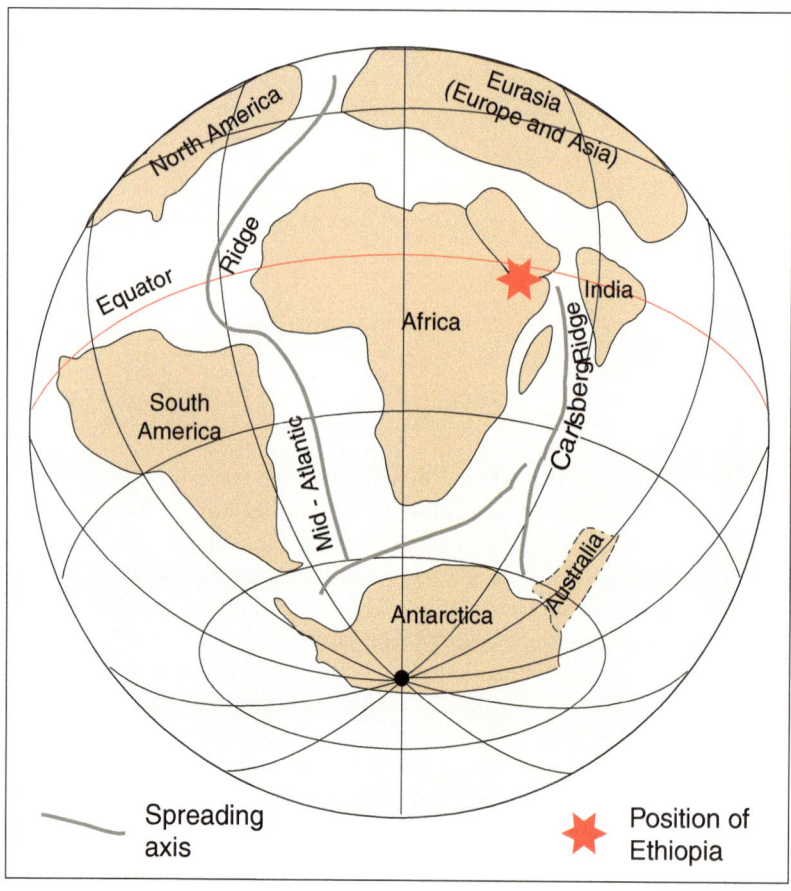

Fig. 7.1 Gondwana at about 60 million years ago

The situation in the region around Ethiopia is shown in more detail in Fig. 7.2. A seaway called the Tethys Sea connected the Atlantic and Indian Oceans, and separated Africa and India from Eurasia to the north. This was becoming narrower because the African and Indian Plates were moving northwards, and its floor was being subducted beneath Eurasia. However, although the African and the Indian Plates were both moving northwards, they were travelling at very different speeds— the Indian Plate at about 18 cm per year and the African Plate at about 3 cm per year. This had caused a rupture between them, known as the Owen Fracture Zone, along

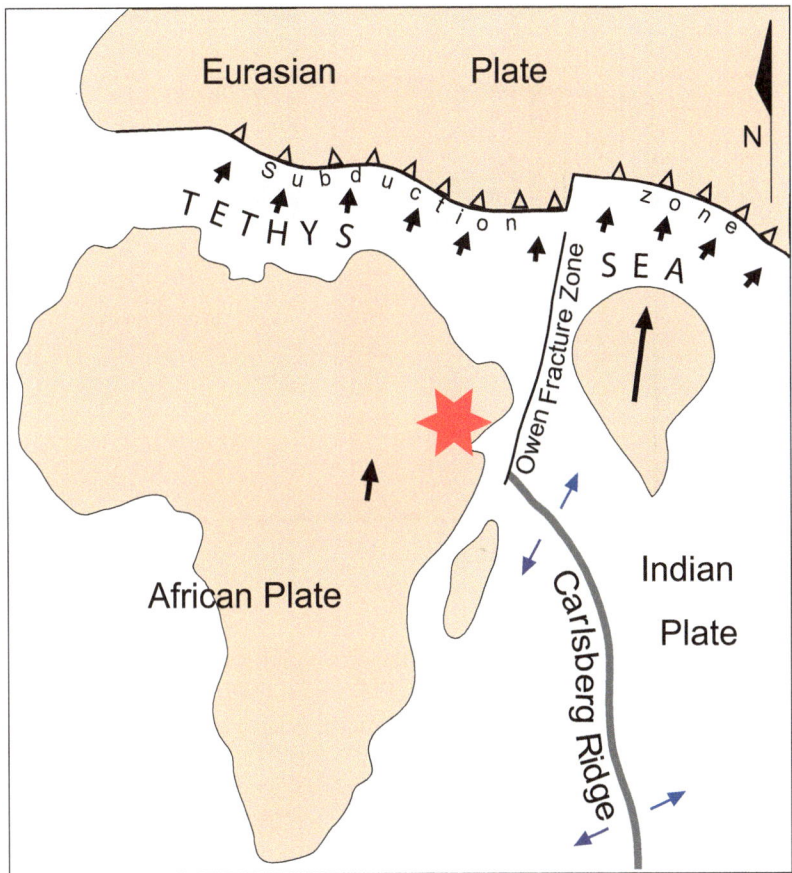

Fig. 7.2 The situation between Africa, India and Eurasia about 60 million years ago. The *black arrows* indicate the direction of plate movement, and the *blue arrows* the direction of sea-floor spreading. The *red star* shows the position of Ethiopia

which the plates were not so much sliding as scraping alongside each other. The Indian Plate was trying to drag the northeastern part of the African Plate, including the Arabian Peninsula, northward, while at the same time the rest of the slower moving African Plate was holding it back. Eventually, something would have to give!

In the meantime, a hot-spot was developing beneath Ethiopia. Hot-spots, more technically referred to as "thermal anomalies", are just what their name implies: abnormally hot localised regions somewhere beneath the earth's surface. They

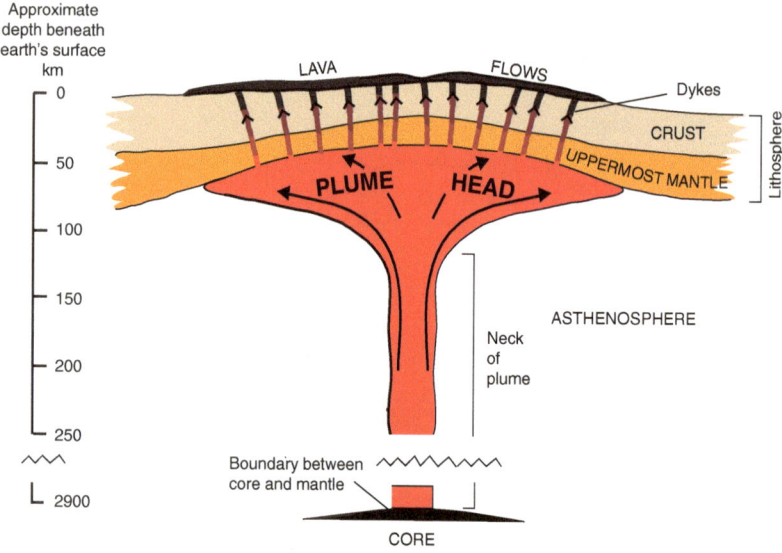

Fig. 7.3 A mantle plume

occur in a number of places around the world and are manifested in volcanic regions such as Hawaii, Iceland and Yellowstone. Their cause is, as with many geological phenomena, disputed. A widely accepted hypothesis is that they are the result of a feature called a mantle plume.

A mantle plume is envisaged as a column of hot material rising from deep within the earth's mantle, possibly from as deep as the boundary between the mantle and the core (Fig. 7.3). The root cause may lie in movements of the very hot, fluid material forming the outer part of the core producing a concentration of heat which acts like a hotplate, heating the mantle immediately above it and causing it to become abnormally hot. This hot material is less dense than the solid mantle surrounding it, and slightly plastic, and slowly moves upward in the form of a narrow neck. As it rises the weight of rock above it becomes less, so that the pressure on it decreases and it begins to melt. When it reaches the base of the rigid lithosphere the partly molten mass spreads out sideways, rather like the mushroom cloud of an atomic bomb. The lithosphere above it is pushed upward and cracks, allowing molten rock to flow out onto the surface.

In the following sections it may sound as though plumes are definite, proven entities but it is important to remember that they are just an idea. Although there are a number of geophysical observations that support their existence, no-one knows for sure that they do actually exist and some researchers don't believe in them at all! However, the hypothesis does help to explain many of the events that follow in the dramatic evolution of the region so let us assume, for want of a better explanation, that it is valid.

The plume, which we will refer to as the Afar Plume, would have begun its life long before any of its effects reached the earth's surface. No-one knows for sure when the hot material might have begun its slow rise upwards, but around 40 million years ago (in the mid-Eocene) its head reached the base of the lithosphere, probably somewhere beneath what is now the western margin of Afar. Here it spread out sideways, and pushed up the overlying lithosphere into a broad dome. This is often referred to as the Afro-Arabian Dome. It extended from western Ethiopia to the Arabian Peninsula but the greatest uplift was beneath Ethiopia herself. As the dome rose, molten rock (lava) poured through cracks in the lithosphere and solidified to form the volcanic rock which today covers the Ethiopian highlands and part of the southwestern Arabian peninsula. Hundreds upon hundreds of lava flows erupted from fissures all over the surface of the rising dome, eventually covering an area of some 600,000 km^2 and piling on top of each other to reach, in some places, thicknesses of more than 2000 m. It is hard to estimate the total volume of lava, as its thickness varies greatly from place to place. One estimate, probably a little on the low side, is 250,000 cu km. Whatever the case—it is a lot of lava! Lavas formed in this way are sometimes called flood lavas (or flood basalts if they are basalt) since they flow over the land surface rather like flooding water would.

These layers of volcanic rock are known as the Trap Series, or Trap Series volcanics. "Trap" comes from an old Swedish word meaning "stairs", because the flows often erode to give a staircase-like appearance (PHOTO 7.1). They consist mainly of basalt, a fine-grained and dark-coloured rock rich in iron and magnesium (see Table 3.1). We shall be looking at the Trap Series in more detail when we visit the Western Highlands in Chap. 12.

After about 29 million years ago eruption of the Trap Series lavas slowed and became more sporadic, but the domed region continued to rise and is still doing so today, reaching spectacular heights. In some places Precambrian and Mesozoic rocks which would not have been much above sea level before all this happened, are today found at elevations of over 2000 m. That, plus the great thickness of the Trap Series on top, means that parts of Ethiopia may have reached heights of 4000 m. And that was not all. Volcanic activity gradually changed its style and, instead of flowing out from cracks and fissures, lavas began erupting through vents to form huge, almost circular volcanoes. Such volcanoes are known as shield

PHOTO 7.1 Trap Series basalt layers rise like a giant staircase in this valley between Gonder and Lalibela, in the highlands of central-northern Ethiopia. The *yellow colour* is due to Meskel flowers which carpet the highland countryside during springtime (2014)

volcanoes, as they are broad in comparison to their height and supposedly resemble flat-lying shields. The ages of the Ethiopian shield volcanoes vary from about 29–10 million years. The best known and most spectacular is the Semien, the highest mountain massif in Ethiopia, which we will visit in Chap. 12.

In this chapter we have rushed through more than 20 million years of volcanic activity in just three paragraphs. In reality of course it all happened very slowly, and years, centuries or even millennia would have elapsed between individual eruptions, even during the peak of activity. Had anyone been living there at the time (and of course it was long before anyone was) they would most likely have been unaware that anything was going on. We will look further into the question of size and frequency of eruptions in Chap. 12. In the meantime we will move on to the next stage in the break-up of Gondwana, and the incipient break-up of Ethiopia herself.

The Break-Up of a Continent, and a Summary of Events

8

The events outlined in the previous chapter: uplift, outpouring of the Trap Series volcanics and the formation of big shield volcanoes, were the precursors of even more dramatic times. Indeed, these were already beginning as the uplift and volcanic activity were taking place. In geology it is seldom possible to place events in a neat time sequence—several things are generally happening at the same time. In this chapter a great deal will be happening, but the maps in Fig. 8.1 should help you to follow the story.

Figure 8.1(i) summarises the situation at about 30 million years ago. Most of Ethiopia and part of the Arabian Peninsula had been pushed upwards into a dome, much of which was covered by the Trap Series volcanics. India, not shown on the map, had collided with Asia sometime between 35 and 50 million years ago but the Tethys Sea still separated Africa and Eurasia. The collision had slowed down the Indian Plate, but it was still moving northward at about 5 cm per year. The African Plate was also moving northwards, but more slowly at about 2 cm per year, so the two plates were still scraping against each other along the Owen Fracture Zone. In the meantime the Indian Ocean spreading axis (the Carlsberg Ridge) was continuing to grow, but had curved around (you can see this if you look back to Fig. 7.2) and was heading toward eastern Africa. At the fracture zone it had become offset sideways a little—and re-named the Sheba Ridge as you can see in Fig. 8.1(ii).

The northeastern part of Africa, however, was in difficulties. The Indian Plate was dragging on it across the Owen Fracture Zone, trying to pull it northwards, but it was still part of the slower-moving African Plate. It was in a kind of plate-tectonic tug of war and something would have to give way. The something that triggered the give was the ever growing Afar Plume.

© Springer International Publishing Switzerland 2016
F.M. Williams, *Understanding Ethiopia*, GeoGuide,
DOI 10.1007/978-3-319-02180-5_8

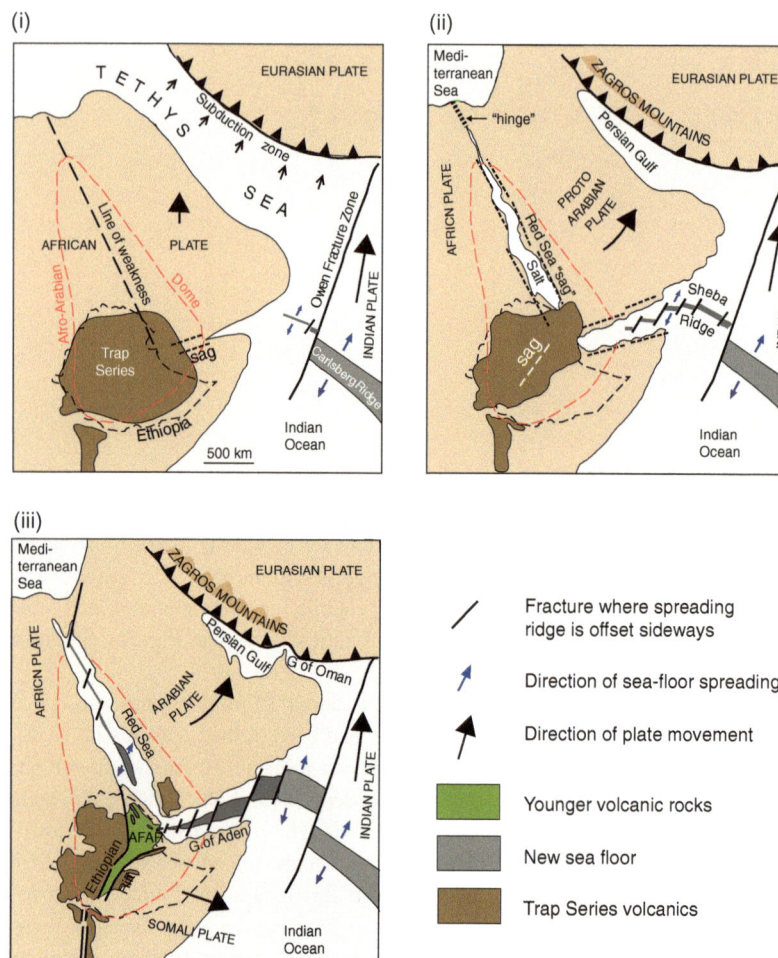

Fig. 8.1 The situation in the Ethiopian region at approximately (**i**) 30 million years ago, (**ii**) 15 million years ago and (**iii**) the present time. Note that everything south of the subduction zone (shown by the *toothed line*), including Africa, is moving northwards as a whole. The *black arrows* show the plate movements in relation to Africa, and do not show this overall northward movement. Diagrams adapted from Bosworth et al. (2005)

As it grew, the plume pushed the overlying lithosphere upwards and outwards, causing it to stretch. In addition, it produced exceptionally hot, plastic mantle material immediately beneath the lithosphere, enabling it to slide. As it stretched and slid, parts of it became thin, sagged and eventually broke apart. The first place to give way was along the future Gulf of Aden, where the newly forming Sheba Ridge was already nosing its way into the continent. By about 18 million years ago it was producing new sea floor between what were to become Somalia and the Arabian peninsula. Thus the Gulf of Aden was born (Fig. 8.1(ii)). Meanwhile the African Plate had made contact with the Eurasian Plate, dividing the Tethys Sea into two: the Mediterranean Sea and the Persian Gulf.

While this was happening events were also under way along the line of the future Red Sea (Fig. 8.1(ii)). This had almost certainly been a zone of weakness since the events of the East African Orogeny more than 500 million years previously. Firstly, possibly as long as 25 million years ago, the weak zone sagged to become an elongated trough, then widened scissor fashion with a "hinge" at its northern end in the region of today's Gulf of Suez. On several occasions the "hinge" broke open, allowing salty sea water to flow in from the Mediterranean Sea. When it closed again the salt water evaporated. In this way thick deposits of salt formed on the floor of the Red Sea trough (Fig. 8.1(ii)).

Although the Red Sea trough continued to widen, and eventually to fracture, a sea-floor spreading axis may not have started to form there until about 5 million years ago. The Gulf of Aden spreading axis (the Sheba Ridge) meanwhile continued to penetrate westward into the African landmass, prising apart Africa and the Arabian Peninsula. As a result of this, and of the scissor-like widening of the Red Sea, the Arabian peninsula was rotating anti-clockwise as it pulled away from Africa. Eventually a spreading axis developed in the Red Sea, beginning about two-thirds of the way down it and growing northwards and southwards. It continues to grow slowly northwards today, and it seems as though it should have continued southwards as well, to meet with the Gulf of Aden spreading axis. However, although there is a narrow seaway, the strait of Bab el Mandeb, connecting the Red Sea and the Gulf of Aden, the two spreading axes are not connected (Fig. 8.1(iii)). Something seems to have got stuck, preventing the Arabian Peninsula from making a clean break from Africa. The key to this "something" lies in Afar, where the two spreading axes terminate and break up into a complex system of faults, fractures, lava fields and ranges of volcanic mountains.

We shall be exploring this fascinating region, and investigating further just what is going in Afar, in Chaps. 21–23.

Sometime between 15 and 10 million years ago, when sea-floor spreading was already well under way in the Gulf of Aden, and a trough had formed along the

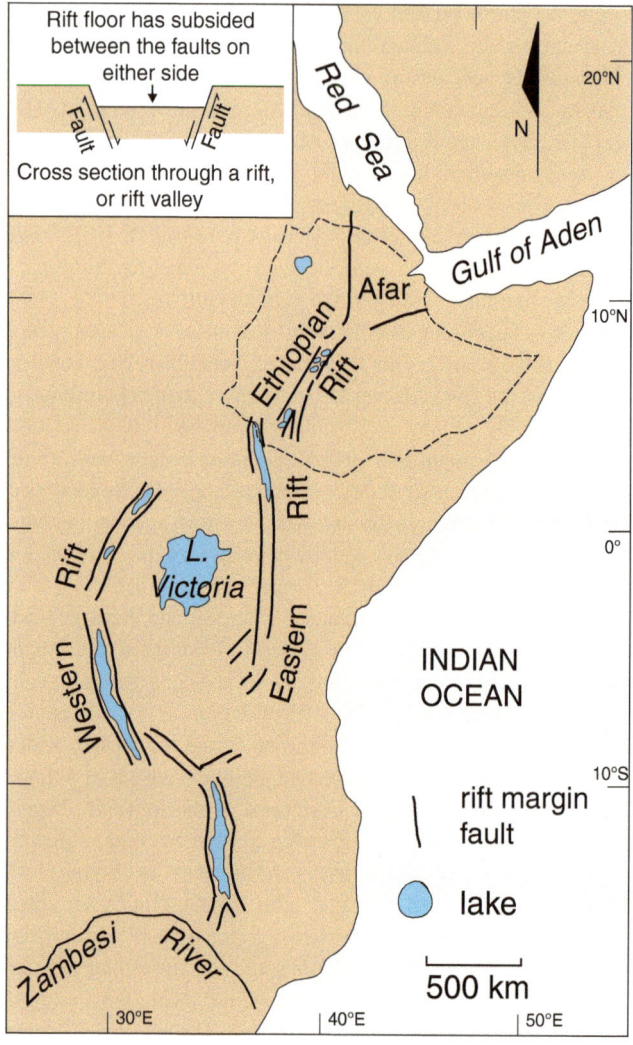

Fig. 8.2 Simplified map of the African Rift System. The *inset* illustrates the structure of a rift, or rift valley

Red Sea, the African continent itself began to split. This started in much the same way as it had in the Red Sea and the Gulf of Aden. From Ethiopia to Mozambique, almost as far as the Zambezi River, strips of land sagged to form troughs, then fractured to form rift valleys. A rift valley (often shortened to simply "rift") is basically a long strip of land which has subsided between two opposite sets of fractures, or faults, as illustrated in the inset to Fig. 8.2. A rift valley is similar to a rifted basin, described in Chap. 6, but generally refers to a narrower and more clearly defined feature. The rift valleys which formed along the eastern side of the African continent are known collectively as the African Rift System. This is shown in simplified form in Fig. 8.2. The northernmost section of it cuts right through Ethiopia and is known as the Ethiopian Rift Valley (often shortened to just Ethiopian Rift). We shall be looking at the African Rift System, rift valleys in general and the Ethiopian Rift Valley in particular in Chaps. 16–19.

Much of the African Rift System, and probably the line of the Red Sea as we saw earlier, follow old structures of the Arabian-Nubian Shield and the Mozambique Belt, along which the ancient continental blocks of Africa, Azania and India had collided during the East African Orogeny. This collision zone had remained as a band of weakness in the earth's lithosphere, rather like the way the two halves of a broken dinner plate, even if glued and pressed firmly together, will still tend to break most readily along the join. The join, or suture, will always be weaker than the rest of the plate. Whatever the origin of the forces pulling Gondwana apart, they are exploiting such zones of weakness.

The Red Sea and the Gulf of Aden are currently widening at a rate of between 1.5 and 2 cm per year. The African Rift System is also widening, but at a much slower rate of less than 0.5 cm per year and it is still, with the exception of Afar, floored entirely by continental crust—there is no sign of sea-floor spreading. It is possible that the Rift System will continue to evolve and become an ocean, in which case the eastern part of the African continent comprising present-day Somalia, Somaliland, the eastern parts of Ethiopia, Kenya and Tanzania, and the northern part of Mozambique, will become a new continent. Alternatively, and perhaps more likely, it will cease to widen and stay much as it is, becoming a defunct or "failed" rift.

Even though they have not completely separated, the regions divided by the Red Sea, the Gulf of Aden and the African Rift System are generally assigned the status of plates: the African (sometimes called Nubian) Plate, the Arabian Plate and the Somali Plate (Fig. 8.1(iii)). The spreading axes of the Red Sea and the Gulf of Aden, and the northernmost section of the African Rift System, all terminate in Afar but don't quite connect with each other. A locality like this, where three plate margins meet or almost meet, is known as a triple junction. If you look back to Fig. 1.2 you

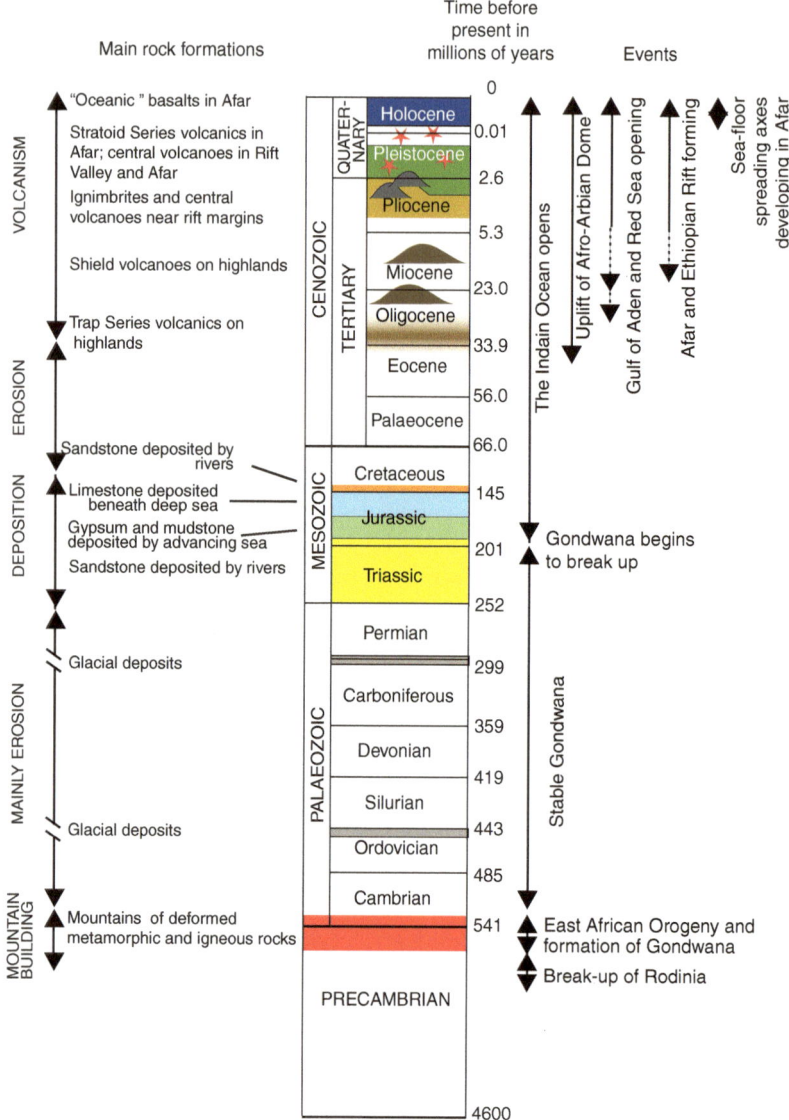

Fig. 8.3 Geological time scale showing the main geological events and corresponding rock formations in Ethiopia

will see that there are a number of triple junctions on the earth, but Afar is unique as it is the only one above sea level—where we can actually observe what is happening.

Most of the volcanic activity in the updomed highland regions had ceased by about 10 million years ago with eruption of the great shield volcanoes. However, volcanic activity was far from over in Ethiopia. If anything it was on the increase, but had shifted from the highlands to the newly forming rift valley and Afar regions as shown by the green areas on Fig. 8.1(iii). This was hardly surprising as it was there that all the new breaks and cracks were forming, allowing lava to penetrate to the surface. In northernmost Afar the most recent lavas are being erupted along volcanic ranges which very much resemble sea-floor spreading axes, and provide an important clue as to what may happen next in this region. We will be returning to this point in Chaps. 25 and 26.

This has completed our brief summary of Ethiopia's geological story. It is now time to begin our journey around the country to observe the manifestations of these events and how they have resulted in the rock formations and the beautiful and varied scenery we see in Ethiopia today.

To help you to locate yourself in geological time, and in the complex sequence of events we have witnessed in the last four chapters, Fig. 8.3 shows again the geological time scale. It is the same as the one that was introduced in Chap. 2, but superimposed on it are the events that were taking place, and the types of rocks being formed, at each stage. I hope that this will help you to visualise, at a glance, the overall story.

The following chapters take the reader on a journey through Ethiopia to observe the geological features and landscapes resulting from the events outlined in the previous pages. To cover the whole country in detail would require a book much larger than this one, but I have tried to include as many as possible of the localities which have a particular geological significance and attraction, and which are most likely to be on a visitor's itinerary. For the non-traveller, or the armchair traveller, I hope that it will convey the variety and beauty of Ethiopia's scenery, and for all readers, an insight into the geological processes which lie behind it.

Before beginning this journey, it is enlightening to take a bird's-eye view of the country. Thanks to modern technology, we are able to do this without taking to the air. Increasingly sophisticated techniques of aerial photography, satellite imagery and digital elevation modelling can provide overhead views which show at a glance what is often not obvious from ground level. Figure III.1 on the next page shows a digital elevation model of Ethiopia and her neighbouring regions. This technique plots land surface elevation (and ocean depth), using data obtained from satellite measurements, in different shades of colour, much like a physiographic map but designed to provide a three-dimensional impression. This kind of picture is very useful for a region such as Ethiopia which has big contrasts in elevation. It is important to note, though, that the colours merely represent elevation and not the actual colour of the terrain. Ethiopia is not brown and grey! The elevation ranges corresponding to the different shades of colour are indicated on the key at top left of the figure.

Looking at the digital elevation model, you should easily be able to pick out the topographic regions that were described in Chap. 4: the Western and Southeastern Highlands, the Rift Valley, Afar and the Ogaden. Many of the features discussed in

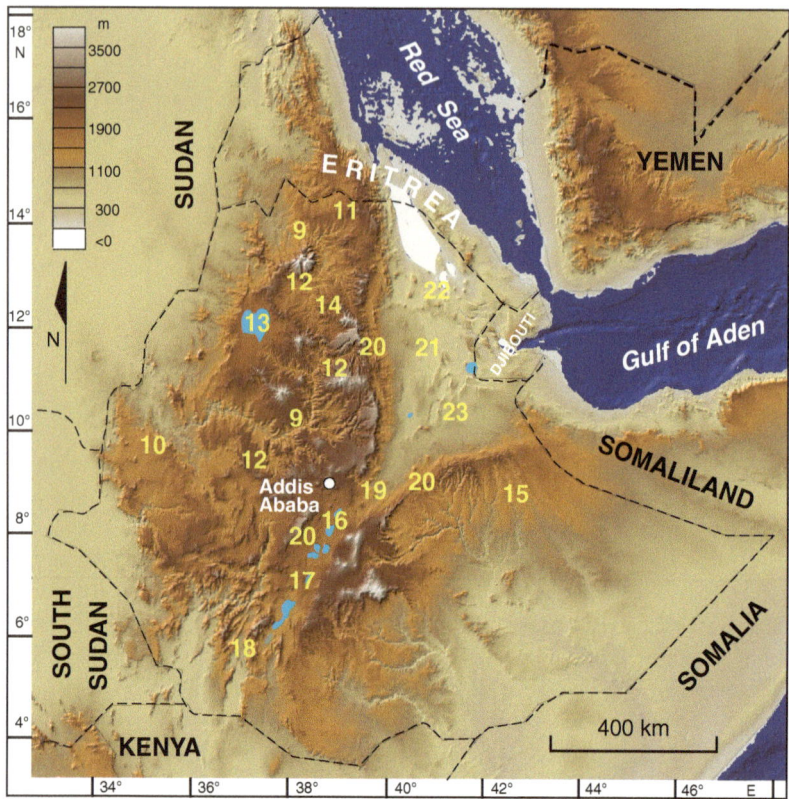

Fig. III.1 Digital elevation model of Ethiopia and her surrounds. The numbers refer to the particular regions and features which will be visited in the following chapters: *9* the gorges of the Blue Nile and Tekeze Rivers (Chap. 9); *10* Western Ethiopia (Chap. 10); *11* Tigray (northern Ethiopia) (Chap. 11); *12* The Western Highlands (Chap. 12); *13* Lake Tana and the Blue Nile (Chap. 13); *14* Lalibela (Chap. 14); *15* The Southeastern Highlands (including the Ogaden) (Chap. 15); *16* The Rift Valley (introduction) (Chap. 16); *17* The central and southern Main Ethiopian Rift (Chap. 17); *18* The broadly rifted region of southern Ethiopia (Chap. 18); *19* The northern Main Ethiopian Rift (Chap. 19); *20* The margins of the Rift Valley and Afar (Chap. 20); *21* Afar (introduction) (Chap. 21); *22* Northern Afar (Chap. 22); *23* Southern Afar (Chap. 23). Note that some *numbers* appear more than once on the figure, as they refer to a broad region or to more than one feature. DEM from GeoMapApp

Chaps. 5–8 can also be observed: for example, the Gulf of Aden spreading axis and how it is penetrating into Afar, and the high escarpments separating Afar from the highlands, and a closer look will reveal more detail such as the big shield volcanoes of the Western Highlands and the deep gorges of the Blue Nile and its tributaries.

Each of Chaps. 9–23 focuses on either a particular region of the country or a geological theme. After taking a vertical view of the geology, revealed by the deep gorges of the Blue Nile and Tekeze Rivers, the chapters follow very roughly a chronological sequence, starting with the oldest and moving forward to more recent events and geological features. It is impossible, though, to adhere rigidly to this plan since almost every region contains rocks and landforms resulting from different periods of geological time.

Although it is preferable to read the chapters in order, it is not essential to do so. I trust that you will be able just to dip in and enjoy!

The Great Gorges: Slices Through Time

9

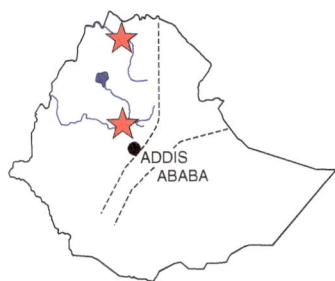

No traveller through Ethiopia can fail to be awed when standing on the lip of one of her great gorges, such as those of the Blue Nile and its tributaries, and gazing down over their steep cliffs and broad terraces to a tiny strip of water shimmering in the depths. These gorges have presented a challenge to travellers, and indeed to any kind of communication, across the highlands of Ethiopia throughout her history. To the geologist, however, they are a boon, as they slice through the great pile of rock formations of which the highlands are built, revealing the geological story which the different layers tell. There are many magnificent gorges in Ethiopia, but those of the Blue Nile and Tekeze Rivers will be described here as they are the most frequently and easily traversed (Fig. 9.1). Between them, they record most of Ethiopia's geological story, at least up to about 25 million years ago after which the focus of events moved to the Rift Valley and Afar.

9.1 The Blue Nile Gorge

The Blue Nile, or Abay, flows from its outlet at Lake Tana and makes a great semicircular loop through the Western Highlands before crossing the plains of Sudan to meet the White Nile at Khartoum. On its way it carves a mighty gorge

© Springer International Publishing Switzerland 2016
F.M. Williams, *Understanding Ethiopia*, GeoGuide,
DOI 10.1007/978-3-319-02180-5_9

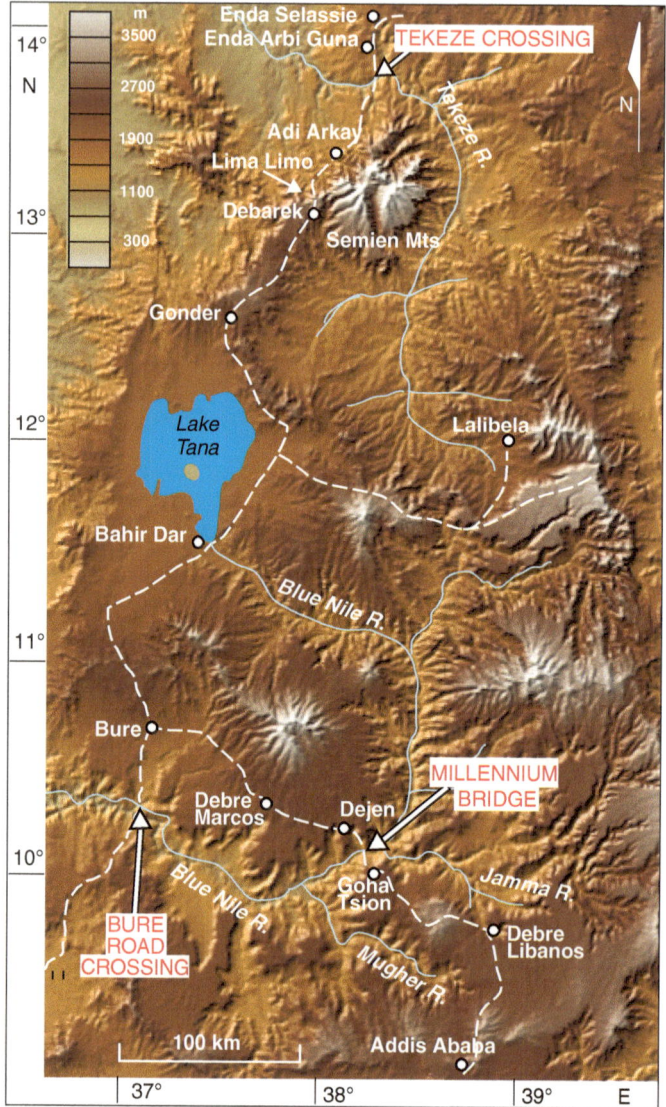

Fig. 9.1 Digital elevation model showing the Blue Nile and Tekeze Rivers and locations where main roads, shown as *white dashed lines*, traverse their gorges. DEM from GeoMapApp

which is sometimes termed the Grand Canyon of Africa. The gorge is over 800 km long and reaches depths of more than 2 km. In Chap. 13 we will look in more detail at how the gorge formed and why the river follows such a circuitous course. In this chapter we will examine the succession of rock formations that it reveals, and see how each relates to the geological story outlined in Chaps. 5–8.

Most of the Blue Nile Gorge is difficult to access, but fortunately a spectacular section of it is traversed by the main road from Addis Ababa to Bahir Dar, which crosses the Blue Nile at the Millennium Bridge (Fig. 9.1). Making this traverse, a traveller passes not only through a kilometre and a half of altitude but also through more than 250 million years of geological time.

Figure 9.2 shows a geological section from the top of the gorge near the town of Goha Tsion, to the Millennium Bridge where the road crosses the Blue Nile River. You will notice that the rock layers are grouped into formations. A formation simply means a group of rock layers that are formed of similar materials and were deposited under a similar set of conditions. If a formation consists of just one rock type the word "formation" is often omitted; the Adigrat Sandstone Formation, for example, is usually just called the Adigrat Sandstone. As noted in Chap. 6, the name of a formation generally refers to the place where it was first described, or where it is particularly well exposed.

Looking across the gorge it is quite easy to distinguish the different formations, as their rocks weather and erode rather differently (PHOTO 9.1).

At the top, the Trap Series volcanics are recognisable by their steep cliffs and the staircase-like profile into which they have been eroded and for which they are named. These are the rocks formed from the lavas which poured out over much of Ethiopia some 30 million years ago as the Afro-Arabian Dome pushed upwards, heralding the break-up of Africa and Arabia. Here the Trap Series rocks are basalt, dark grey to black in colour and arranged in thick horizontal layers, each of which represents an individual flow of lava. In this part of the Blue Nile Gorge the Trap Series reaches a total thickness of around 300 m but, as noted in Chap. 7, there are places where it is as much as 2000 m thick. Many of the flows show spectacular columnar jointing, which can be seen in road cuts along the steep descent into the gorge (PHOTO 9.2). These pillar-like structures form as the hot lava cools and contracts, rather like the way wet mud shrinks and cracks as it dries out.

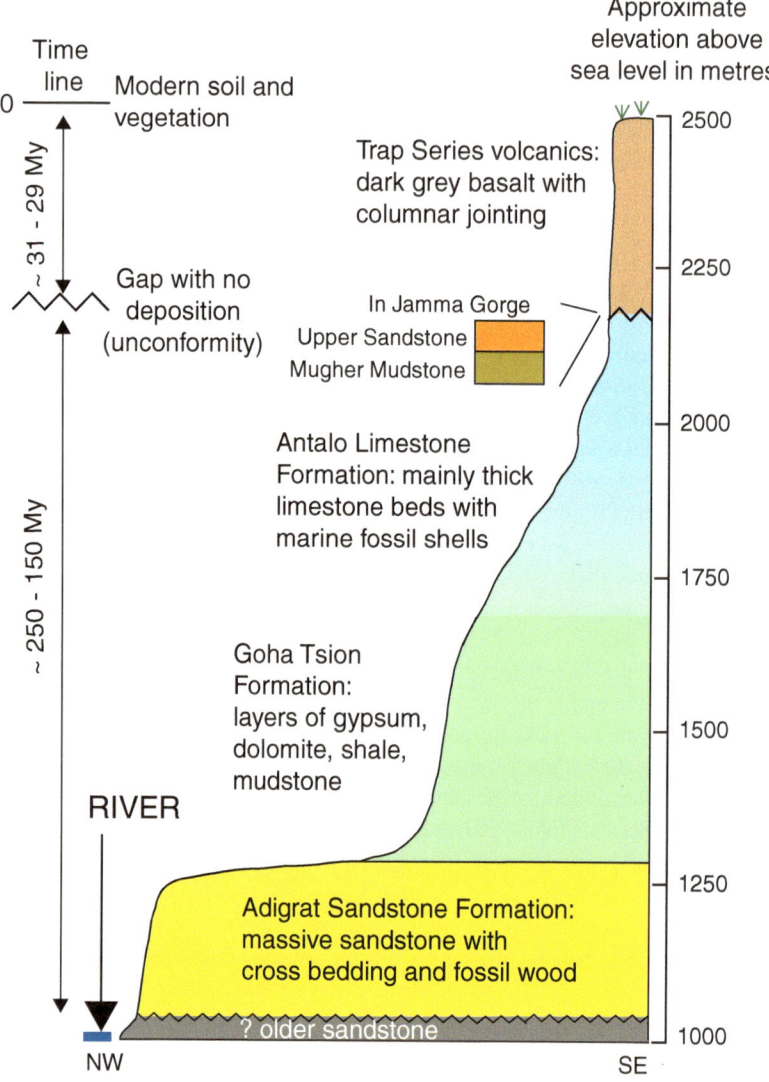

Fig. 9.2 Geological section through the Blue Nile Gorge, between Goha Tsion and the Millennium Bridge river crossing

PHOTO 9.1   View over the Blue Nile Gorge, from about half-way down on the descent from Goha Tsion. The step-like cliffs at the top of the gorge are formed by Trap Series volcanics, underlain by Antalo Limestone which shows as a white-ish layer. The sloping surface is formed of limestone, dolomite, mudstone and gypsum of the Goha Tsion Formation, and the cliffs directly above the river of Adigrat Sandstone. (2012)

Immediately beneath the layers of Trap Series basalt, the type of rock changes abruptly to thick layers of white limestone (PHOTO 9.3). This is the Antalo Limestone Formation (named after the town of Hintalo in Tigray), deposited beneath the sea which moved northwestward across Ethiopia during the Jurassic period until, around 150 million years ago, it covered a large part of the country. Limestone consists of calcium carbonate, and may be formed by precipitation of this chemical compound from the seawater, by accumulation of the shells and other remains of organisms that lived in the sea or on the sea bottom or, as in the case of the Antalo Limestone, by both. The limestone layers contain numerous fossil shells, particularly those of brachiopods (a kind of bi-valved shellfish) and gastropods (a snail-like creature) (PHOTO 9.4). Enterprising local children have realised a profitable occupation in selling these to visitors, so there is no need to go fossicking!

PHOTO 9.2   Columnar jointing in Trap Series basalt near the top of the Blue Nile Gorge (2012). Photo courtesy Andrew Dakin

Although the Trap Series basalts lie directly on top of the limestone, there is a time gap of over 100 million years between their deposition. Such a time gap is called an unconformity. On geological cross sections, such as Fig. 9.2, unconformities are usually represented by a wavy or zig-zag line. To find out what happened during this time we must look elsewhere, but we will leave that till later in this chapter.

In its lower layers the Antalo Limestone becomes less pure and white, and the beds are thinner. Layers of mud, fine sand and gypsum (a form of hydrated calcium sulphate) appear among the limestone beds and gradually become more prominent. The limestone itself becomes a more magnesium-rich form known as dolomite. The profile of the gorge also changes, as the mud and gypsum erode to a more gradual slope than the stepped cliffs formed by the massive limestone and basalt. These layers of mud, sand, gypsum and dolomite are collectively known as the

PHOTO 9.3 Cliff of Antalo Limestone in the Blue Nile Gorge. The lower layers are finely bedded and grade into the Goha Tsion Formation. (2012)

Goha Tsion Formation, after the town at the top of the gorge (PHOTO 9.5). Gypsum is a mineral formed by evaporation of calcium-rich water, and dolomite is characteristically deposited in warm, shallow sea water. We can therefore envisage a situation of pools and mudflats, alternately drying out and being renewed by outgoing and incoming tides, and building up thick deposits of mud, gypsum and dolomite as the sea slowly advanced and deepened.

Almost at the bottom of the gorge, prominent and unmistakable, are buttress-like cliffs of red sandstone (PHOTO 9.6). Although the Blue Nile Gorge is far from Adigrat in northern Ethiopia, this sandstone is assigned to the Adigrat Sandstone Formation on account of its similarity in composition, age and the manner in which it was deposited. It is built from loads of sand which were carried by rivers during the Triassic period, between about 250 and 200 million years ago,

PHOTO 9.4   Fossil brachiopod shells from the Blue Nile Gorge (2011)

into the shallow rifted basins which heralded the breakup of Gondwana. The composition of the sand shows that it was eroded, as would be expected, from Precambrian basement rocks. Several things can also be deduced from the detailed structure of the sandstone layers. A feature known as cross-bedding, where finer beds run diagonally across the main, horizontal ones, indicates that the rivers which deposited the sand flowed from the northwest. The size and varying orientations of the cross-beds suggest that the rivers were braided and meandering, and sometimes spread out to form large deltas. Fragments of fossil wood tell us that they flowed across a forested land (PHOTO 9.7), and fossil pollen spores confirm that the age of the sandstone is Triassic.

You may remember from Chap. 6 that the first sedimentary rocks to be deposited on the Precambrian basement were tillite and sandstone, laid down during glacial episodes of the Palaeozoic era. Tillite has not been found in the Blue

PHOTO 9.5 Layers of gypsum, mud and dolomite of the Goha Tsion Formation (2014) (Geological hammer 30 cm long for scale)

Nile Gorge; however, it is thought that a white layer of sandstone close to the river may have been deposited around that time.

There is another formation missing in this section of the Blue Nile Gorge—the Upper Sandstone which, as we saw in Chap. 6, was deposited by rivers as the Jurassic sea retreated. Either it was not deposited at this spot, or was eroded away before the lavas of the Trap Series erupted. It is seldom possible to find at one locality a sequence which contains all the formations laid down in a region. Generally, one has to search further and then piece things together.

It is not necessary to look far for the Upper Sandstone, as it is present in some of the nearby tributary gorges such as that of the Jamma River near Debre Libanos (PHOTO 9.8), and in fact it is sometimes termed the Debre Libanos Sandstone. The lower part of this sandstone is very fine-grained and muddy, and has been given its own name, the Mugher Mudstone, after the nearby Mugher River gorge

PHOTO 9.6   Buttress-like cliffs of Adigrat Sandstone near the bottom of the Blue Nile Gorge. The Millennium Bridge can be seen at *bottom left* (2014)

where it is also found. It is a very interesting formation, as fossilised fragments of fish and reptiles have been found in it, including teeth of ancestral crocodiles and even those which may have belonged to a relative of the great dinosaur *Tyrannosaurus rex.*

The Upper Sandstone and the Mugher Mudstone fill only a tiny fraction of the gap, or unconformity, between the Antalo Limestone and Trap Series basalts. After the Upper Sandstone was deposited, there was still over 100 million years to go before the Trap Series lavas flowed over the land.

At the Millennium Bridge crossing our journey back into deep geological time is curtailed, since here the river has not cut deep enough to reach the rocks of the Precambrian basement. It does not do so until about 40 km downstream of the bridge, and the closest place where they can be seen in the gorge in the comfort of

PHOTO 9.7 Fossil tree trunk in the Adigrat Sandstone, near the bottom of the Blue Nile Gorge (2012)

a vehicle is about 130 km to the west, where the road from Bure to Nekemte crosses the river (Fig. 9.1). At that crossing the river has cut deep into highly metamorphosed rocks of the Mozambique Belt.

9.2 The Tekeze Gorge

The Tekeze River rises in the mountains near Lalibela (Fig. 9.1). It flows northward before curving around the Semien Mountains and then follows a northwestward course into Sudan, where it joins the Atbara River and then the Nile. North of the Semien Mountains, the river and its tributaries have cut away part of the mountain itself as well as slicing through the underlying rock layers. Just how this happened will be discussed further in Chap. 12. The important point for this

PHOTO 9.8   The gorge of the Jamma River near Debre Libanos. The cliff directly below the stepped layers of Trap Series basalt, and above the sloping surface, is formed of Upper Sandstone (2010)

chapter is that it has created another informative and spectacular transect of Ethiopia's geological past. Like the Blue Nile, it is conveniently traversed by a main road, that which runs from Gonder to Enda Selassie (Shire) and the famed historic sites of northern Ethiopia.

The road section between Debarek and the Tekeze River must be one of the most scenic of Ethiopia's many scenic roads. Shortly after passing through Debarek, which is situated on the western shoulder of the Semien Mountains, it plunges almost without warning into a precipitous and in places hair-raising descent of an escarpment known as Lima Limo. I have been reliably assured that to date (2015) no vehicle has gone over the edge! This is followed by a spectacular traverse through the spires and pinnacles of the Semien foothills (which we will visit in Chap. 12) before descending to the valley proper of the Tekeze River (PHOTO 9.9).

PHOTO 9.9   The Tekeze River, winding its way through Precambrian basement rocks (2014)

The geological section through which the road passes is shown in Fig. 9.3. You can see at once that it is very different from the Blue Nile section. The gorge cuts well into the Precambrian basement rocks, there is only a very thin layer of sandstone, no limestone and a huge pile of volcanic rocks.

These volcanic rocks consist of Trap Series volcanics like those we saw in the Blue Nile gorge, but here overlain by lavas of the Semien shield volcano. It is hard to tell, though, where one ends and the other begins. As we will see in Chap. 12, there is in fact very little time gap between them. Studies of small differences in the structure and chemical composition of the flows have suggested that the contact between them occurs at about 2700 m, but I have exercised caution and marked it with a query on Fig. 9.3.

The sandstone seen in the Tekeze gorge is the same formation, Adigrat Sandstone, that we saw in the Blue Nile gorge but here it is much thinner. This could be because there was nothing to protect it from erosion during the 200 million years or so that elapsed before the Trap Series lavas flowed over it, or perhaps it was just

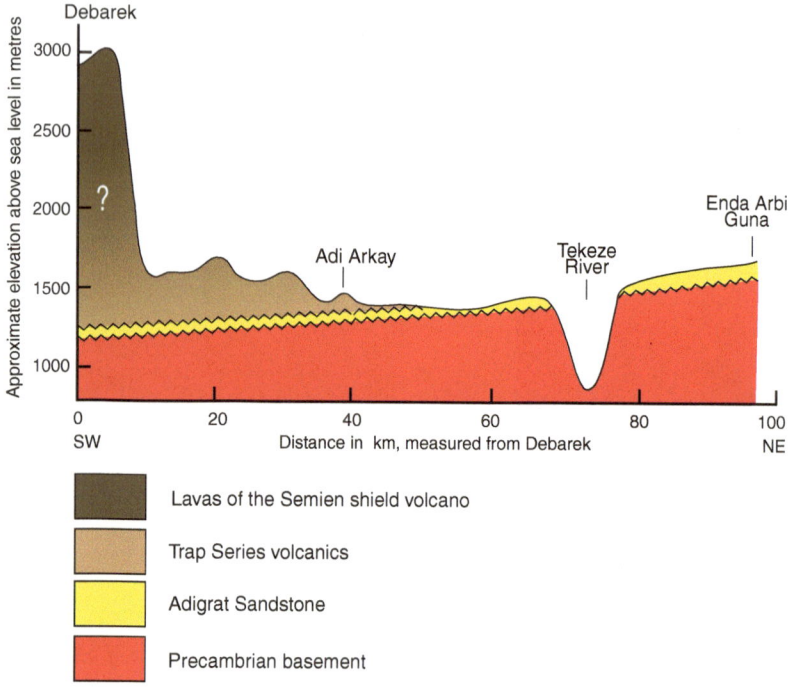

Fig. 9.3 Geological cross section through the Tekeze Gorge between Debarek and Enda Arbi Guna. The *question mark* indicates uncertainty as to where the boundary is between the lava flows of the Trap Series, and those of the Semien shield volcano

thinner in the first place. The absence of any Jurassic limestone, not only here but anywhere in northwestern Ethiopia, suggests that the Jurassic Sea never reached this far.

The Adigrat Sandstone here lies directly on the Precambrian basement; no Palaeozoic deposits are seen. There was however a long time gap, about 300 million years, between the formation of the basement and the deposition of the sandstone. During this period the basement rocks were worn to a flat plain, and anything that might have been deposited on them was eroded away. The basement rocks here, and in northern Ethiopia in general, belong to the Arabian-Nubian Shield and were originally volcanic rocks and ocean sediments uplifted, squeezed and metamorphosed during the East African Orogeny. In this part of the Tekeze Gorge they consist of a rather soft and crumbly slate which is very easily eroded: firstly because

PHOTO 9.10 Slate in road cutting on the southern side of the Takeze River. The slate easily breaks along its cleavage planes and slides onto the road. If you look carefully you can see a geological hammer for scale (2014)

water readily penetrates between the slatey layers, or cleavage planes, and secondly because the layers easily slide over one another (PHOTO 9.10). This makes it difficult for the highway authorities who are attempting to widen the road through the gorge, since the sliding rocks cause road-cuts to collapse almost as soon as they are made.

By combining the information revealed by these two slices through time, we have seen how many of the major events in Ethiopia's geological history are recorded in the layers of rock that were deposited. Not all the story is there, as you can see by referring to the geological time scale in Fig. 8.3. Many important events occurred later, and some events, such as the glaciations of the Palaeozoic era, are not recorded at these localities. For the geologist who loves to travel (and which of them does not?) this is a compelling reason to explore elsewhere to fill in the missing pieces.

Although western Ethiopia is not a common destination for visitors, it is a good place to begin our geological tour since much evidence for the early stages of Ethiopia's geological story can be found here. It is also home to some of Ethiopia's most interesting and attractive Precambrian rocks.

In this chapter western Ethiopia is considered to be roughly the region west, northwest and southwest of the town of Nekemte, shown in the geological map in Fig. 10.1. The map may look a little complicated, but you can see that the dominant colours are light brown, dark red and orange-red. Light brown represents the Trap Series volcanic rocks which cover most of the Ethiopian highlands. They may originally have extended across western Ethiopia as well but have been mostly eroded away. Red represents Precambrian rocks: dark red those which belong to the Mozambique Belt and orange-red those which belong to the Arabian-Nubian Shield. All of these Precambrian rocks are either metamorphic, meaning that they were subjected to conditions of high temperature and pressure during the East African Orogeny, or are igneous intrusions formed from molten rock material which solidified beneath the earth's surface.

You will recall from Chap. 5 that the Arabian-Nubian Shield represents the remains of ocean floor, ocean sediments and volcanic island arcs that were squeezed, crushed and metamorphosed as the African and Indian cratons, and the continental

© Springer International Publishing Switzerland 2016
F.M. Williams, *Understanding Ethiopia*, GeoGuide,
DOI 10.1007/978-3-319-02180-5_10

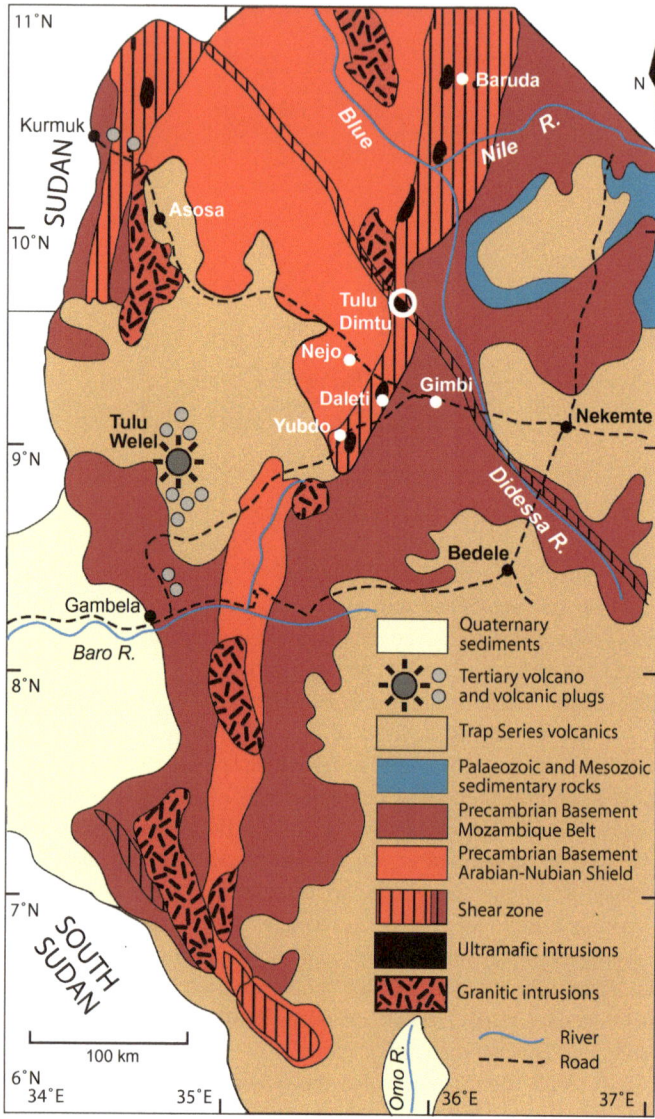

Fig. 10.1 Geological map of western Ethiopia. Modified from Alemu and Hailu (2013)

fragment of Azania, came together during the East African Orogeny. The rocks thus formed are generally rich in dark-coloured minerals which contain magnesium and iron, and are termed mafic (**ma**gnesium, plus **f**errum, the Latin word for iron). The rocks of the Mozambique Belt are richer in light elements such as potassium and sodium, and since they contain a higher proportion of silica (silicon dioxide) they are termed silicic. This is one of the few parts of Ethiopia where these two families of Precambrian rocks can be observed side by side, and it has been a challenge for geologists to sort them out and determine just where one ends and the other begins.

The sections marked with vertical stripes on the map are shear zones. Shear zones are produced when blocks of crust scrape alongside each other, rather like a car accident in which two vehicles sideswipe each other rather than colliding head on. The hill known as Tulu Dimtu, indicated by a white circle on the map, is particularly interesting as it is located where two shear zones intersect each other. The first shear zone, running NNE-SSW has itself been sheared in a NW-SE direction. Rocks metamorphosed in shear zones are often formed of platey or stringy minerals as a result of being stretched and dragged by the sideways movement (PHOTO 10.1).

The collision which resulted in the East African Orogen was extremely complex. Since the whole collision process lasted for more than 100 million years there was time for a lot to happen and it is not easy to work out the details, particularly as it all happened over 500 million years ago. Returning to our car accident analogy, it is sometimes quite difficult to work out from the positions of the cars and the type of damage suffered exactly what happened during the collision. It is vastly more difficult in the case of colliding continents! Rock layers were compressed, bent into folds, thrust over each other, then perhaps folded a second or even a third time, completely altering their original arrangement. Heat and pressure changed their mineral composition, and sometimes they were heated so much that they partly melted, resulting in a swirly looking rock called migmatite ("mixed" rock) (PHOTO 10.2).

To add to the complexity, large quantities of magma (molten rock) of varying composition pushed their way into the system before, during and after the collision event, solidifying to form masses of coarse-grained rock. As explained in Chap. 5, these are known as intrusions. The intrusions shown on the map in speckled red are termed granitoids. Granitoid simply means a granite-like rock—it may not be granite itself but any similarly coarse-grained rock. The intrusions which are coloured black on the map all occur within the shear zones and are described as

PHOTO 10.1 Serpentinite in a quarry near to Tulu Dimtu mountain. This serpentinite is formed of the fibrous green mineral serpentine and the soft platey mineral talc, both of which are silicate minerals containing iron and magnesium. This was originally an ultramafic rock such as dunite (see PHOTO 10.3) which has been altered by pressure and temperature in a shear zone. Ethiopian one *birr coin* for scale (2014)

ultramafic. Ultramafic means very (**ultra**) rich in **ma**gnesium and iron (**ferrum**) (PHOTO 10.3). These rocks have a composition very similar to that of the earth's mantle, and are actually the result of part of the mantle itself melting and rising toward the surface.

A rather unusual rock, or rather group of rocks, found here is termed ophiolite. This name comes from a Greek word *ophios* meaning "snake" and *lithos* meaning "rock". The greenish colour of some ophiolite rocks is apparently associated with that of snakes (the same concept applies to the serpentinite rock shown in PHOTO 10.1). Ophiolite is a mix of basalt (fine-grained mafic rock), gabbro (coarse grained mafic rock) and dunite or pyroxenite (both ultramafic rocks)—all

PHOTO 10.2 Migmatite in a quarry east of Gimbi. Migmatite ("mixed rock") forms when metamorphism has been so intense that some of the rock has actually melted, or at least become very plastic and able to flow. This has caused the light and dark minerals in the rock to segregate into distinct bands. You can almost see the movement of the molten material in the swirls (2014)

metamorphosed. Ophiolites are important because, by careful examination of them, geologists are able to distinguish which parts of the ocean floor or island arc system they were before they were metamorphosed. They thus provide very good clues in piecing together the events occurring during the collision.

Regions of metamorphic rocks such as this, especially where there are shear zones and intrusions, are good places to look for mineral deposits. The cracks and fractures formed in shear zones as the rocks drag against each other provide passages for watery solutions produced during the metamorphic process, and from intrusions as they solidify. Elements such as copper, gold, silver, lead, platinum, sulphur and others which do not fit into the crystal structure of the common

PHOTO 10.3 Serpentinised dunite from a quarry near Daliti, west of Gimbi. Dunite is an ultramafic rock composed almost entirely of the mineral olivine (an iron magnesium silicate). The melt from which this rock solidified must have come from the earth's mantle. As it came closer to the earth's surface the olivine became partly altered to fibrous serpentine (the green streaks) (2014)

rock-forming minerals become concentrated in this fluid material, which generally also contains a lot of excess silica (silicon dioxide) and sometimes potassium. Because it is watery it crystallizes slowly so that very large grains of quartz and feldspar can grow. Rocks formed in this way are called pegmatites and are easy to recognise on account of their big grains (PHOTO 10.4). They are a good place to look for minerals such as gold, and gemstones such as topaz, beryl or tourmaline.

Although a number of valuable minerals, such as platinum, gold, nickel and copper, occur in the rocks of western Ethiopia, the deposits have thus far been little exploited. Platinum has been mined at Yubdo, and local people pan for gold in some of the rivers in the region, collecting gold flakes that have been washed out of

PHOTO 10.4 Granite gneiss at the Didessa River road crossing. This is a metamorphic rock formed mainly of quartz and potassium feldspar. Gneiss has a typically banded structure, and this one is called a granite gneiss because it has a similar composition to granite. Adjacent to the *hammer* is a small vein of pegmatite, formed of coarse grains of quartz and potassium feldspar. Pegmatite grains can grow much bigger than this, occasionally to more than 10 m across! (2014)

pegmatites upstream. The potential for larger-scale gold mining is, at the time of writing, still at the exploration stage.

Minerals which contain iron and magnesium (such as pyroxene, hornblende and olivine) tend to break down rather easily when they come into contact with water and with the atmosphere, so that the rocks of which they are constituents are easily weathered and eroded. Much of the landscape of this region is therefore one of rounded hills and meandering river valleys (PHOTO 10.5), a contrast to the dramatic gorges and sharp precipices of the Western Highlands, and to the knobbly hills characteristic of the Precambrian parts of southeastern and southern Ethiopia

PHOTO 10.5 Scenery north of Gimbi, looking towards Tulu Dimtu hill (2014)

which we will visit in Chaps. 15 and 18. You might indeed wonder why it has not worn down altogether, to a flat plain. The answer is that the region, which is situated close to the western extremity of the Afro-Arabian Dome (see Chap. 7), has been slowly rising for the past 30 million years or so, continually countering the rate at which is being eroded.

Almost all the basement rocks of western Ethiopia are between 850 and 500 million years old. As explained in Chap. 5, the very few ages that are older than this have been measured on tiny crystals of zircon which are scattered through the rock. Zircon (zirconium silicate) is a very resistant mineral, not easily altered by heat and pressure. Zircon crystals continue to grow around their edges when subjected to new temperatures and pressures (PHOTO 10.6), but their interior core may remain unchanged and be as old as when it first crystallised from a magma in one of the ancient cratons. The successive growth layers of such zircons carry

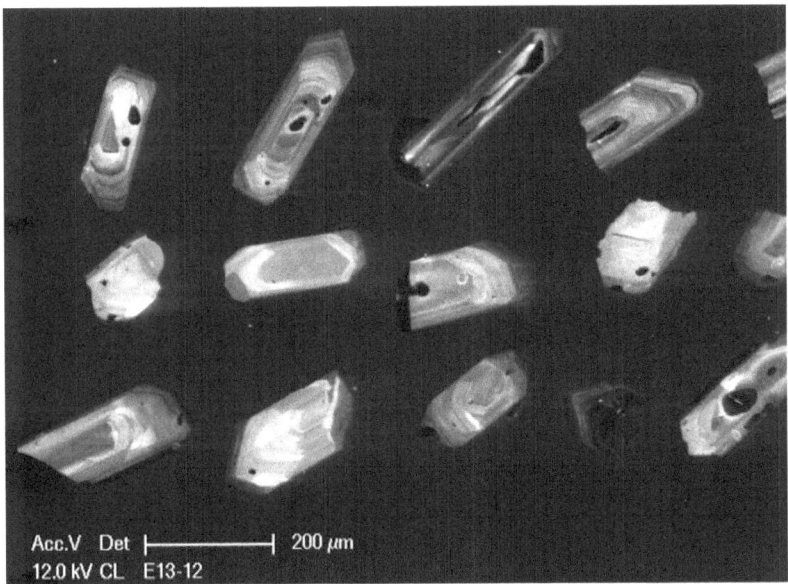

PHOTO 10.6   Zircon crystals from western Ethiopia. Layers have been added around the edges as the crystals grew and temperature and pressure conditions varied, but the central part has remained unchanged. The crystals are tiny: 20 microns, indicated by the scale is equal to 0.0002 mm (2014). Photo courtesy Morgan Blades

information about the whole of their geological history since they first formed, often before even Rodinia itself came into existence. Zircons provide the key to unravelling the entire history of the earth's crust—but that is another story!

Some of the rocks in this region are much younger. Travellers who venture close to the border with Sudan, on the roads to Kurmuk or Gambela, may notice some oddly shaped dome-like or pointed hills, rising from an otherwise flat plain (PHOTO 10.7). They are volcanic plugs, formed of material which has solidified within the vent of a small volcano and then been slowly pushed upwards, or become exposed when the rest of the volcano eroded away. We shall be looking at volcanic plugs in more detail in Chap. 11 when we visit the region of Adua in northern Ethiopia. The large volcano Tulu Welel is formed of trachyte (a moderately silica-rich rock; see Table 3.1) and is probably, like the plugs, between

PHOTO 10.7 Volcanic plugs between Asosa and Kurmuk (2014). Photo courtesy John Foden

about 10 and 7 million years old. This volcano has some interesting stories attached to it, ranging from being the original King Solomon's Mine—the source of the wealth of that famous king—to being the home of the Devil! Tulu Welel will feature again briefly in our story, in Chap. 12.

Landforms, Monuments and Hidden Churches of Tigray

Although Tigray, the northernmost *kilil* (administrative region) of Ethiopia, is topographically part of the Western Highlands, it is distinct in almost every other respect. Firstly, its climate is much drier. It has its own language, dress, hairstyles, music, dance and traditions. The stone walls and houses, skilfully built of local stone that blends into the landscape, are characteristically Tigrayan, and the further you penetrate into the region the more strikingly its scenery differs from that of the highlands to the south.

The geological map in Fig. 11.1 gives a clue as to why this is so. Southwest of a diagonal line which partly follows the course of the Tekeze River, the map is almost entirely brown (Trap Series volcanics) with a few sedimentary rocks (yellow) exposed in river gorges; northeast of this line it is mostly red (Precambrian basement rocks). Unlike the highlands to the south, those of Tigray are not covered by thick layers of volcanic rock, and the scattered patches that do occur are thin, less than 400 m thick compared to 1000 m and more over much of the region further south. Here we are reaching the northern extremity of the Trap Series lava flows.

© Springer International Publishing Switzerland 2016
F.M. Williams, *Understanding Ethiopia*, GeoGuide,
DOI 10.1007/978-3-319-02180-5_11

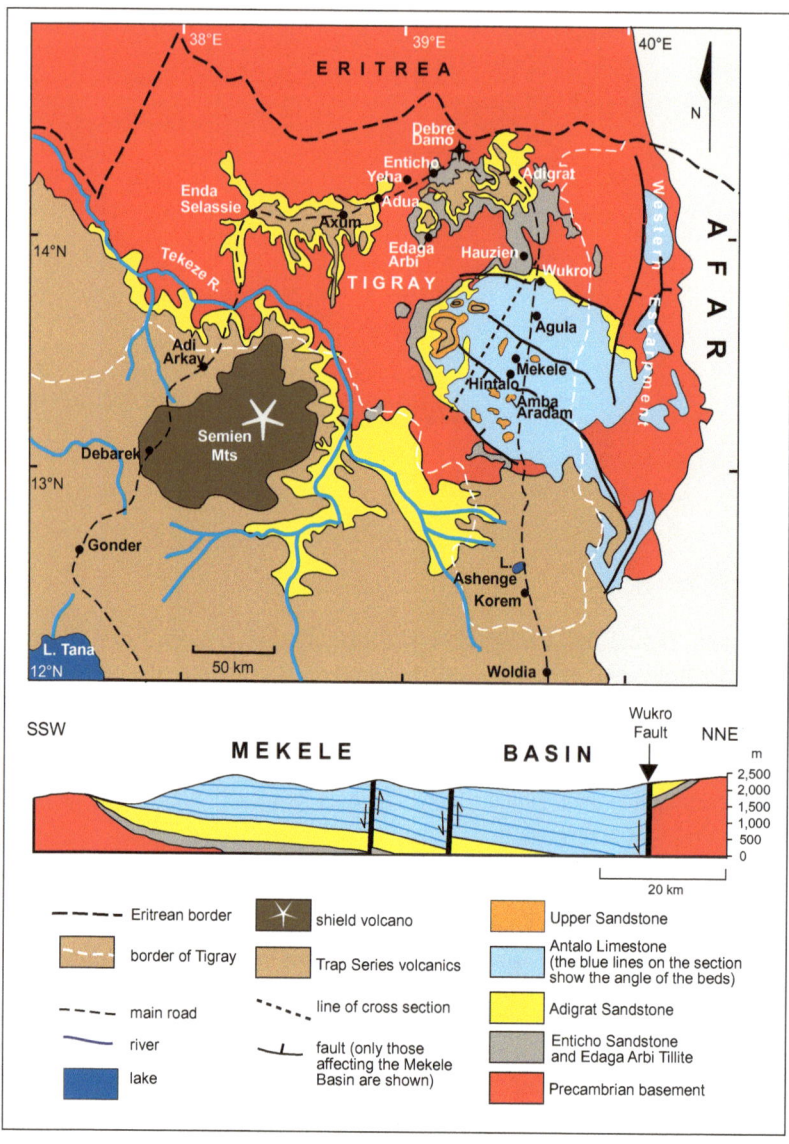

Fig. 11.1 Geological map of the highland region of Tigray, and part of the highlands to the south. *Below* the map is a geological cross section of the Mekele Basin, along the *black dotted line* shown on the map. The map and cross section are modified from Geological map of Ethiopia 1:2,000,000 2nd Ed. (1996) and Geological map of Mekele 1:250,000 (1971), Geological Survey of Ethiopia

The Precambrian basement in Tigray belongs to the Arabian-Nubian Shield, and is formed mainly of metamorphosed volcanic rocks and ocean sediments, with numerous granite intrusions (these are not distinguished on the map). The overlying sedimentary rocks and Trap Series volcanics have been mostly eroded away, and in the far north and in Eritrea may never have been deposited. There are three main regions where they do occur: the large area coloured blue around Mekele, the sandstone region (mostly coloured grey) between Hauzien and Adigrat, and the ridge of sandstone, partly capped by volcanic rocks, which extends from Adigrat to Enda Selassie. Each of these regions has features of particular interest, and we will look at them in turn.

11.1   The Mekele Basin

The conspicuous oval area to right of centre on the geological map, coloured in blue and with the city of Mekele at its centre, is known as the Mekele Basin. It is just that—a basin-like depression in the Precambrian basement, filled with Palaeozoic and Mesozoic sedimentary rocks. Just why there should be a basin here, and what caused it to form, is not entirely understood. Its northeastern side is bounded by faults, whereas its southwestern side is partly a fault and partly a slope. The cross section shown below the geological map in Fig. 11.1 gives the general picture. The eastern edge of the basin is cut off along the western escarpment of Afar. Possibly it started life as a rifted basin, like those described in Chap. 6, though this would not account for its distinctive oval shape.

Before the basin formed, much of this area had been covered by the glacial deposits of the Palaeozoic era. Tillite, composed of fragments of material carried and dumped by glaciers (see Chap. 6), and sandstone formed from sand deposited in their meltwater lakes and streams, overlie the Precambrian basement rocks and line the floor of the basin. They are known respectively as the Edaga Arbi Tillite and the Enticho Sandstone. You can see the places after which they are named on Fig. 11.1, and a photo of the Edaga Arbi Tillite in PHOTO 6.1. When the basin formed, by whatever means, it provided a deep embayment into which Mesozoic rivers flowed, depositing layers of sand which covered the older sandstone and tillite. This became the same Adigrat Sandstone Formation which we met in the Blue Nile Gorge, and you can see on Fig. 11.1 that the town of Adigrat, after which it is named, is located just a little way to the north.

PHOTO 11.1 Surface karst landform in Antalo limestone in the Mekele Basin. These features are caused by solution of the limestone (2011)

Following this the basin was flooded by the Jurassic sea which moved in from the southeast, depositing thick layers of limestone. Limestone is by far the most abundant rock seen throughout the basin since, apart from around the edges, it covers everything else. It is the same Antalo Limestone Formation that we encountered in the Blue Nile Gorge, and the town of Hintalo after which it is named is situated near the centre of the Mekele Basin. Limestone is rather easily dissolved by water, and in places here this has resulted in a landscape of strange fluted pillars and ravines, known as surface karst (PHOTO 11.1). The same limestone is found on the Danakil Alps (which we will come to in Chap. 22) on the eastern side of Afar, and even further away in Yemen. Remembering that in the Mesozoic era these would all have been connected, it is quite possible that the cut-off southeastern edge of the Mekele Basin is in Yemen! Outside of the Mekele Basin, apart from some appearances along the western escarpment of Afar, there is no more Jurassic limestone in Tigray. As well as approaching the limit of the flows of Trap Series volcanics, we have reached the northwestern extent of the Jurassic sea.

As the sea retreated, rivers flowed over the limestone and deposited layers of sand—the Upper Sandstone. Between the limestone and the sandstone is a fine-grained, thinly-bedded rock known as the Agula Shale. Finally, as the Afro-Arabian Dome rose during the Tertiary period, Trap Series basalt erupted to cover the sandstone, though probably only thinly. As a result of the uplift the limestone, originally deposited at sea level, now stands at an elevation of over 2000 m.

The rock layers overlying the limestone in the Mekele Basin have been almost entirely eroded away. The remnants of them stand out as small, steep-sided and flat-topped hills. Such hills are known geologically as mesas (originally from a Latin word *mensa* meaning "table"), but in Ethiopia they are called ambas, a term which signifies "mountain fortress". The secret to their survival is that the upper layer of sandstone was baked by the lava which flowed over it, forming a hard cap which protected the softer rocks below. Some of the mesas are still topped by a thin layer of basalt. The best known of the mesas of the Mekele Basin is Amba Aradam, which stands prominently near the main road south of Mekele. It was the site of a famous battle, the Battle of Amba Aradam, in 1936, when the advancing Italian army defeated Ethiopian forces defending the amba. In this case geology worked against the Ethiopians, as the Italian troops encircled the amba and, despite their brave resistance, the Ethiopians were trapped.

The Upper Sandstone is sometimes referred to as the Amba Aradam Sandstone, particularly in this region. As we saw in Chap. 9, in the Blue Nile region it is sometimes called the Debre Libanos Sandstone. It quite frequently happens in geology that the same formation is given different names in different places and for this reason I will stay with the more general term Upper Sandstone.

11.2 Hauzien to Adigrat, and the Hidden Churches of Tigray

Tigray's geological make-up has resulted in some surprising landscapes. North of the Mekele Basin one comes upon a scene which is more reminiscent of the Wild West than of Ethiopia—a chaotic landscape of pinnacles and mesas carved out of

PHOTO 11.2 Scenery to the south of the Wukro to Hauzien road. The flat plain in the middle ground is underlain by Precambrian basement rocks. The mesas and pillars, formed of Enticho and Adigrat Sandstones, are erosional remnants, the last survivors of a sandstone plateau which once covered the basement and which has now been almost entirely eroded away (2012)

white and red sandstone (PHOTO 11.2). These are the remnants of what was once a sandstone plateau, perhaps topped by basalt, which has now been eroded away with only a few stalwart outposts remaining. Much of the area has been worn right down to the Precambrian basement, which forms a level plain on which the remnants of overlying Enticho and Adigrat Sandstones stand (PHOTO 11.3). A few million years hence, those too will be gone.

It is not only a landscape of impressive beauty, but also one of great cultural, religious and historical significance, for tucked away among the pillars and mesas are over a hundred ancient churches. These are not the prominent edifices that westerners are accustomed to think of as churches, but are hidden away in caves or

PHOTO 11.3 Contact between the Precambrian basement and Enticho Sandstone near Gheralta Lodge, Hauzien region. The basement rock is a strongly foliated (banded) gneiss, which appears *greyish-white* in the photograph; the sandstone is *red-brown* and blocky. There is a time gap of over 100 million years between the formation of the basement rocks and the deposition of the sandstone (2012)

carved into the rock, and are often difficult to access. The little church of Abuna Yemata Guh, for example, is located about a third of the way up the pillar near the right hand side of PHOTO 11.2, and to reach it involves a vertiginous cliff ascent (PHOTO 11.4) followed by negotiation of a hair-raisingly narrow ledge (PHOTO 11.5). Although this may deter visitors who are inclined to vertigo, the beautifully painted interior of the church is well worth the effort (PHOTO 11.6). These extraordinary churches have been painstakingly sculpted, often with beautiful and elaborate interiors of columns and arches carved from the rock (PHOTO 11.7), and decorated with the colourful and evocative paintings that, in

PHOTO 11.4 A cliff of
Enticho Sandstone, the first
part of the ascent to Abuna
Yemata Guh church,
Hauzien region (1973)

PHOTO 11.5 The entrance
to Abuna Yemata Guh is
through the crack at the end
of this narrow ledge. To the
left of the ledge is a sheer
drop of some 250 m (1973)

PHOTO 11.6 The beautifully painted interior of Abuna Yemata Guh makes the nerve-racking approach worthwhile. This ceiling picture portrays nine of the twelve apostles. The remaining three are painted on the church walls. The paintings in this church are in near perfect condition as its roof is overlain by around 500 m of sandstone (1973)

PHOTO 11.7 Carved pillars and cupola of Mikael Kurara church on the Gheralta ridge, southwest of Hauzien. The painting on the *right hand* pillar shows Roman soldiers gleefully crucifying Christ. In traditional Ethiopian paintings bad people are shown in profile, good people full face. Paintings lower on the pillars, and the sandstone itself, are being worn away by worshippers touching and kissing them (2012)

PHOTO 11.8 The Gheralta ridge, formed of Enticho Sandstone, southwest of Hauzien. Several churches are located on this ridge, whose steep sides appear impossible to scale (2012)

every Ethiopian church, tell the stories of the Bible and of Ethiopian saints, and warn graphically of the consequences of a sinful life and the rewards of a good one.

Why there are so many churches, why they are hidden away and difficult to reach, who built them and when and why, are questions for historians. Suffice it to say here that they are very old. Local guides tell that they were built by Abreha[1] and Atsbeha, twin brother kings who were converted to Christianity in Ethiopia (then the Kingdom of Axum) during the 4th century AD. It is probable that many, though not all, of them date from this time, and owe their state of preservation both to the dry climate of the region and the thick layers of sandstone that protect them.

Whatever the reason may have been for creating hidden and inaccessible churches, geologically a better place could not have been chosen. The dissected

---

[1]Abreha may correspond to the historic king whose "throne name" was Ezana.

PHOTO 11.9 A basalt dyke has pushed through a crack in the sandstone of the Gheralta ridge, widening and filling it, and was then eroded away leaving a narrow passageway. Basalt columns from the eroded dyke provide a rough staircase. The walls of the passage are covered by a smooth, hard, black coating where the hot lava of the dyke baked the sandstone as it intruded. This dyke is on the way to Mariam Korkor church (2012). Photo courtesy Andrew Dakin

landscape is naturally full of nooks, crevices, steep precipices and hidden cavities, and the sandstone is soft enough to be hollowed out, either to enlarge what may have been an existing cave or create a new one, and to sculpt the interior decorations. In some cases geology even assists in providing a route to otherwise inaccessible churches. For example, some 35 churches are located near Hauzien, on a long sandstone ridge called Gheralta whose sides are near vertical (PHOTO 11.8). Ascent looks impossible. But it turns out that geology has provided secret staircases! Basalt lava has, in a few places, pushed its way through the sandstone, forming a narrow intrusion known as a dyke. The basalt has then weathered more rapidly than the adjacent sandstone to provide not only a

PHOTO 11.10 The church of Mikael Kurara, located near the *top* of the Gheralta ridge is carved into the sandstone and concealed behind the *white* façade. The church is accessed via an eroded basalt dyke similar to that shown in PHOTO 11.9, followed by a steep cliff climb up a bare rock surface. The interior of this church is shown in PHOTO 11.7 (2012)

passageway through the cliff-side, but also a rough staircase of horizontally jointed columns (PHOTO 11.9). At least two of the Gheralta churches, Mariam Korkor and Mikael Kurara (PHOTO 11.10) are accessed in this way. The basalt has even provided material for the church bells (PHOTO 11.11). These eroded dykes support the idea that the ridge may have once been capped by basalt flows, fed by the lava rising through them.

PHOTO 11.11 The bells of Mikael Kurara church are made from pieces of columnar-jointed basalt taken from the access dyke (2012)

11.3 Adua and Axum: Plugs, Domes and Obelisks

Between the towns of Adigrat and Enda Selassie is a long ridge of sandstone, partly capped by basalt which is probably the northernmost remnant of the Trap Series volcanics (Fig. 11.1). As well as containing some unusual landscapes, this ridge is home to a number of Ethiopia's most important historic sites.

On the northern edge of the ridge, about 20 km northeast of Enticho, is the monastery of Debre Damo, a reclusive settlement of monks and their acolytes dating from the 6th century AD (PHOTO 11.12). Situated on top of an impressive amba surrounded by vertical cliffs, the only means of reaching the monastery is to be pulled up by rope, an adventure unfortunately not available to this author as only people and even animals of the male gender are allowed access.

About half way along the ridge, in the region of Adua and Yeha, there is a break where the sandstone and basalt have been eroded away. Here a strange landscape

PHOTO 11.12 Debre Damo monastery is situated on *top* of this amba, formed of Enticho Sandstone. The only means of access is to be pulled up by rope. Photo courtesy Robert Neil Munro

emerges, of pointed hills, steep sided domes and rocky spires. Some are isolated, others are aligned to form bumpy ridges (PHOTO 11.13). They are commonly referred to as the Adua plugs, and are formed of trachyte (see Table 3.1) and a more unusual volcanic rock called phonolite. Phonolite is similar to trachyte but contains less silica, so that a low-silica mineral called nepheline forms in place of some of the feldspar. The name phonolite derives from the Greek *phone-os*, meaning "voice" or "sound", as some phonolites give a ringing sound when struck.

The reason why these odd looking hills are called plugs is because they, or at least some of them, are just that—a lump of lava that has solidified inside the vent of a volcano and plugged it up. The plug has then been exposed, either by the volcano around it eroding away, or by being pushed up like a piston by pressure from below (Fig. 11.2). Returning to our earlier toothpaste analogy—this is like having a very old tube of toothpaste in which the paste has become stiff and hard inside the neck. If you squeeze the tube hard enough from below, the hard stuff is

PHOTO 11.13 A ridge of plugs and volcanic hills near Adua (2011)

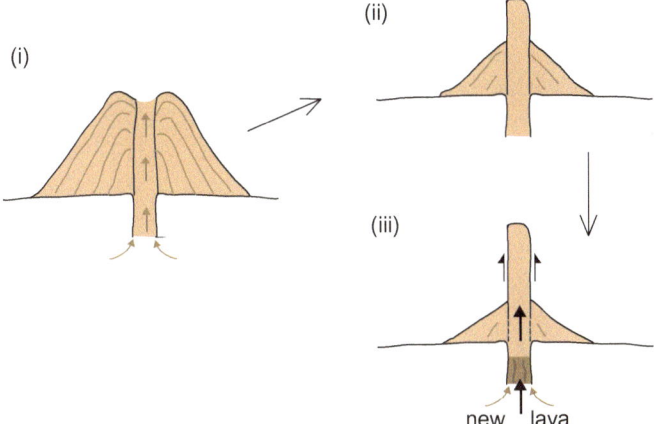

Fig. 11.2 How the Adua plugs may have formed. (**i**) Lava is extruded and forms a steep sided volcanic hill surrounding its vent. When volcanic activity ceases, the whole edifice solidifies. (**ii**) The surrounding material is eroded away, leaving the more resistant material that filled the vent. (**iii**) Volcanic activity recommences. New lava pushes up the material blocking the vent. The plug becomes taller. The hills around Adua may represent all stages between (**i**) and (**iii**)

pushed up. Although some of the Adua hills may have formed in this way, others may not be true plugs but simply steep-sided volcanic cones or domes, like stage (i) in Fig. 11.2.

Adua and its plugs and domes played an essential part in Ethiopia's history, for here the army of Emperor Menelik II defeated the Italian would-be colonialists at the famous Battle of Adua in 1896. Part of the reason for the Italians' humiliating defeat was that they became confused by the hills, which not only look rather alike but in some cases have similar names. One division of the Italian army took up its position on the wrong plug, considerably weakening the Italian defence and assisting the Ethiopians, who were also far better organised, armed and determined than the Italian command had foreseen, to achieve a decisive victory.

Although I have not been able to find any records of the Adua plugs themselves having been dated, some of them have pushed their way through basalts that are around 20 million years old, so they must be younger than this. They are generally considered to be Pliocene in age. Similar plugs and domes are found along the same ridge to the west, and others are dotted over many parts of Ethiopia. We met some in Chap. 10, and will see more in later chapters of this book. In general, they appear to be associated with the latest eruptions of the Trap Series volcanics. Nowhere else, however, is there such a great concentration of them as there is around Adua and Yeha.

Yeha is the remnant of Ethiopia's oldest city. It was established at least 2800 years ago as the capital of an ancient empire called Damot, before the better-known Axumites came on the scene. It is located in a valley surrounded by plugs and steep-sided domes, and the protection they afforded could be the reason for selection of the site for such an important city. It could not have been that they provided desirable building stone, for the tombs, temple and dwellings that are preserved at Yeha are built almost entirely of sandstone blocks that would had to have been transported from several kilometres away. Only in a few places was material from the nearby hills used in the construction. It was too hard to cut into suitably sized blocks, unlike the more easily worked sandstone.

A similar rock type to that of the Adua hills occurs in the region of Axum, a little to the west, but takes a rather different form. Instead of steep-sided domes and pointed hills it occurs as low, flat-topped domes, up to two and a half kilometres across. The rock is also different in texture from that of the plugs—it is the coarse-grained equivalent of trachyte known as syenite (see Table 3.1), or nepheline syenite if nepheline is present. These syenite domes are very important, for it is from them that the rock for the famous Axum obelisks was quarried.

PHOTO 11.14   The 21 m tall obelisk at Axum (2007). Photo courtesy Emile Farhi

There are dozens of obelisks, or stelae,[2] around Axum, all made of this type of rock and erected around the 4th century AD when the kingdom of Axum, which flourished between about 100 BC and 700 AD, was at its height. The tallest and by far the most impressive of the obelisks stand in a group close to the town (PHOTO 11.14). Six of these are over 15 m high,[3] each one sculpted from a single block of stone. The tallest of all was 33 m high, but this one has fallen down, possibly when the people were attempting to erect it. The second tallest, 25 m high, was taken to Rome by the Italians during their occupation of Ethiopia, but was returned in 2005 and has since been re-erected. The third tallest, 21 m high above its basal plinth, has stood upright, albeit at a slightly tilted angle, since its installation some 1700 years ago.

Figure 11.3 shows a geological sketch map of Axum, and Fig. 11.4 illustrates how the syenite domes might have formed and why they are so suitable for

---

[2]The terms obelisk and stele are often used interchangeably when referring to the monuments at Axum. Strictly speaking an obelisk is a four-sided column with a pyramidal top. The tallest of the Axum monuments are therefore correctly referred to as obelisks, whereas the smaller ones, often with broad, curved tops, would fall into the category of stelae.

[3]Reported heights of the obelisks vary by up to 3 m either way! This may be because some estimates include the plinths on which the obelisks are erected, while others do not

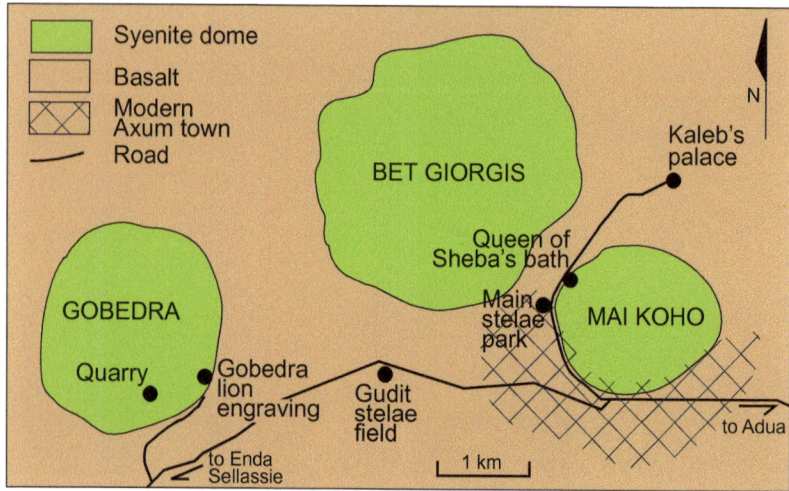

Fig. 11.3 Geological sketch map of Axum, showing the syenite domes from which the obelisks were quarried. *Modified from Asrat et al. (2008)*

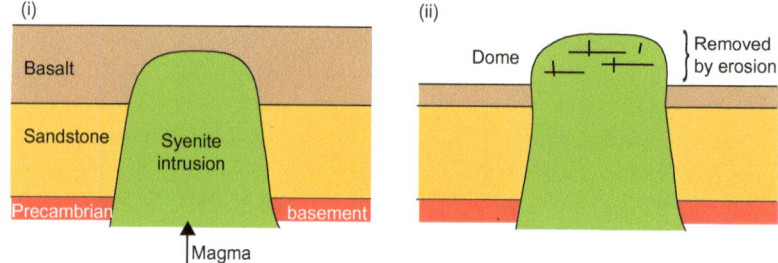

Fig. 11.4 Formation of the syenite domes at Axum. (**i**) Magma intrudes through the overlying rocks but solidifies before it reaches the surface. (**ii**) The overlying rock is eroded, revealing the top of the dome and releasing pressure on it so that the rock expands a little and cracks, producing elongated slabs separated by weaker joints

quarrying. Magma pushed its way through the layers of sandstone and basalt but solidified before it reached the surface. As the overlying basalt eroded away, the upper part of the dome became exposed. Removal of the weight of overlying

PHOTO 11.15 Quarry on Gobedra Hill, Axum. Notice how natural joints separate long slabs of rock, ideal for carving into obelisks (2011)

material relieved pressure on the rock so that it was able to expand or "bounce back" a little, and as it did so it cracked. The structure of the rock was such that horizontal cracks were long and closely spaced, while vertical ones were much more widely spaced, resulting in long slabs of rock separated by weaker joints—a shape perfect for quarrying and dressing into tall, slim obelisks (PHOTO 11.15). The tallest were quarried from the dome called Gobedra (or Gobo Dura), where the slabs are longest. On the nearby dome called Bet Giorgis the vertical joints are closer together enabling only shorter slabs to be quarried. How the enormous blocks of stone were transported and erected remains a matter for speculation.

The obelisks themselves are erected on basalt bedrock which is overlain by clayey soil and loose rock fragments. The basalt is quite weathered, and it is probable that the obelisks, or at least the tallest ones, have a foundation excavated several metres into the rock.

In the next chapter we will move southwards into the heart of the Western Highlands, where the sandstones, limestones and Precambrian basement rocks which give Tigray its distinctive scenery are deeply buried beneath thick piles of lava and the huge volcanoes which sit atop them.

The Western Highlands: Lava Flows and Great Volcanoes

For much of Ethiopia's history the Western Highlands were, essentially, Ethiopia. The Ethiopia we know today, with her present political boundaries, is quite a recent phenomenon. Although Ethiopia is mentioned numerous times in the Bible,[1] as far back as the book of Genesis, the land referred to was not today's Ethiopia but a somewhat vaguely defined region to the south of Egypt. It is often identified with the somewhat later kingdom of Merowe in what is now northern Sudan. Prior to the time of the Emperor Menelik II, who finally incorporated surrounding lands and established the current boundaries of the country in the late 19th century (more or less, as there has since been a fair amount of squabbling over some of them), the only region of present-day Ethiopia that had any kind of coherence was the mountain mass of the Western Highlands, generally known as Abyssinia. Even that was barely a unified country but rather an assortment of disparate princedoms whose rulers were frequently at war with each other.

During medieval times a legend spread through Europe, of a king who ruled over a Christian nation lost among the Muslims and pagans of the Orient. This king, known as Prester John, reputedly presided over a realm full of riches and

---

[1]"Ethiopia" is the Greek translation of the Hebrew name "Cush", which is used in the original Hebrew text of the Old Testament. Modern translations of the Old Testament tend to use "Cush" and "Ethiopia" interchangeably. There is, however, some dispute as to whether "Cush" and "Ethiopia" refer to the same country. In any case, neither corresponds geographically to today's Ethiopia.

© Springer International Publishing Switzerland 2016
F.M. Williams, *Understanding Ethiopia*, GeoGuide,
DOI 10.1007/978-3-319-02180-5_12

strange creatures. The name of Ethiopia became linked to this legend, although few people in Europe had the remotest idea of where Ethiopia was, and the name was sometimes used synonymously with India to mean a vaguely defined land somewhere far away. Whether or not there is any truth in the legend of Prester John, the image of a Christian fortress is an appropriate one. Ethiopia's Western Highlands with their great mountain massifs, formidable gorges and plunging precipices, have long presented a deterrent to would-be invaders and to visitors both friendly and unwelcome. They have provided refuges for beleaguered monarchs, safe locations for churches and monasteries, and natural prisons for undesirable visitors and for anyone who might pose a threat to the monarchy (PHOTO 12.1). Most importantly, they have assisted Ethiopia to preserve her unique flavour of Christianity and traditional way of life.

A digital elevation model of the Western Highlands is shown in Fig. 12.1. They, together with the Southeastern Highlands which we will visit in Chap. 15, are the

PHOTO 12.1 The mountain fortress of Mekdela, where Emperor Tewodros imprisoned the British Ambassador Charles Duncan Cameron and a number of other foreigners between 1864 and 1868 (2012). Photo courtesy Andrew Dakin

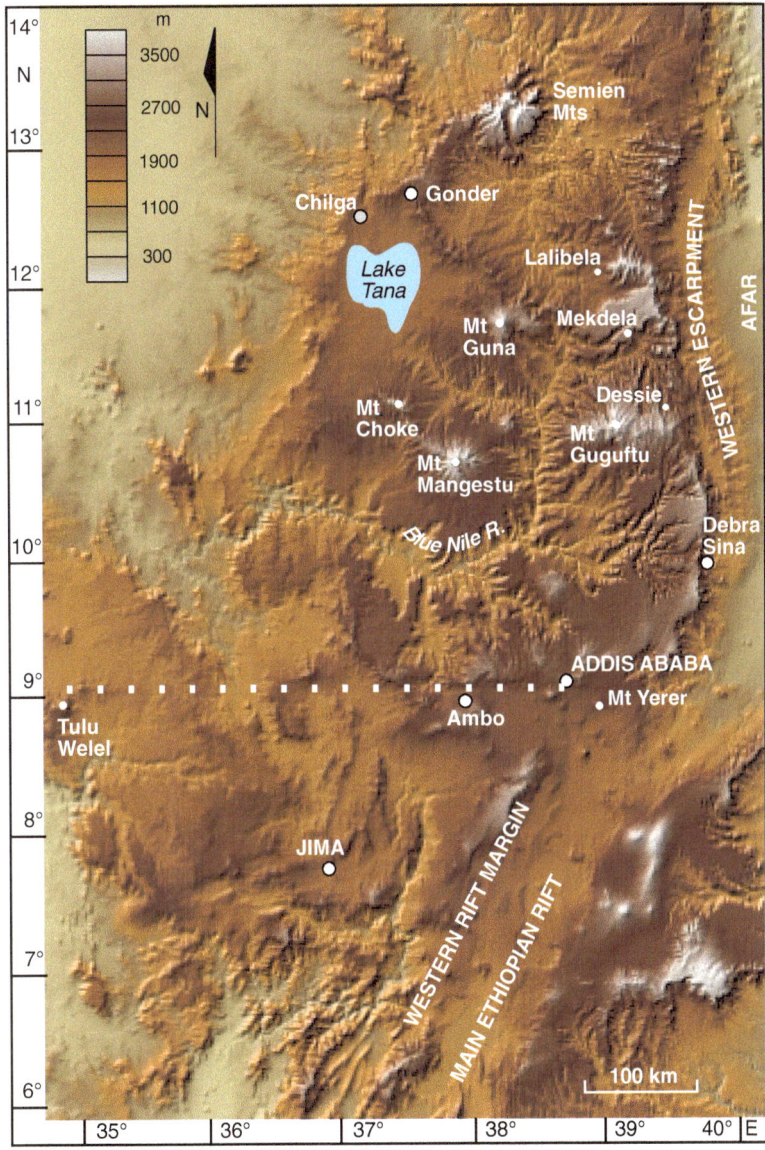

Fig. 12.1 Digital elevation model of the Western Highlands of Ethiopia. Note that the highlands continue into Eritrea, but this is not shown on the figure. The *dotted line* extending westwards from Addis Ababa is the Yerer-Tulu Welel Volcanotectonic Lineament, which is explained in the text. DEM from GeoMapApp

result of two important geological events which were outlined in Chap. 7: the uplift of the Afro-Arabian Dome due to hot material rising as a great plume from the earth's mantle, and the eruption of molten rock through fissures and vents to form a thick pile of lava flows surmounting the dome. Which of these events happened first or whether, as seems most likely, they happened concurrently, is uncertain but whatever the case, the result was a high plateau which may have reached elevations of 4000 m and more above sea level.

12.1   The Trap Series

The pile of lavas erupted over the Afro-Arabian Dome, and which covers most of the Western Highlands and extends into Yemen, is known as the Trap Series as explained in Chap. 7. The lavas have been subdivided into groups depending upon when and where they were erupted, and according to their composition, but here we will simply group them all as Trap Series. They were erupted over quite a long period of time, from about 45 million years ago in the southern part of the Western Highlands to about 15 million years ago in a few localities, and consist of a variety of rock types, from basalt to rhyolite and ignimbrite. However, the most prolific outpouring occurred during a (geologically) very short interval, between 31 and 29 million years ago. These flows were mainly basalts which erupted through fissures, and though these are now buried beneath the lava pile, the conduits that carried the molten rock from below can occasionally be seen cutting through the pile of flows as dykes (PHOTO 12.2). Hundreds of flows were produced, individual ones perhaps spreading over distances of tens or hundreds of kilometres.[2] Sometimes, during the time gaps between the flows, soils had time to develop and can be seen as reddish layers separating the flows.

Following this intense episode of volcanic activity there was, in most places, a break for some 4 or 5 million years. During this time very thick soils had time to

---

[2]It would be interesting to know how frequently eruptions took place during this period. To determine this, though straightforward in principle, is not so easy in practice. It is difficult to follow an individual flow physically in order to determine its size since obstacles such as vegetation, soil cover, overlying flows and deep gorges get in the way, and it has to be tracked by making detailed measurements of its chemical composition. This has not yet been done in Ethiopia, and in very few places elsewhere. Based on the very sparse measurements available, an average volume of 10 cu km for an individual flow seems not unreasonable as an extremely ballpark figure. If half the estimated total of 250,000 cu km of Trap Series lavas were erupted during the 2 million years of maximum activity, a flow of this volume would be required every 150 years or so, somewhere in the country.

PHOTO 12.2 Dyke through weathered basalt flows near Debre Sina. The lava rising through a dyke such as this would have poured out as a flow when it reached the surface. Notice the columnar jointing at right angles to the dyke (2014)

develop on the basalts, and the land became covered by forests, grassland and lakes. Streams carried sand, mud and plant material into the lakes and thick piles of sediment formed, reaching more than 100 m in some places. The forests and grassland, together with the animals that lived in them and the sediments deposited in the lakes, were subsequently buried when volcanic activity recommenced. Over time the fine, clayey sediment altered to a fine form of silica called chert, and the buried plant material was gradually compressed and became lignite, a kind of impure coal.

Because they occur between the lavas of the Trap Series these deposits are known as intertrappean sediments, and they can be seen in numerous places over the Western Highlands. A particularly good example can be seen near Chilga, north of Lake Tana (PHOTO 12.3). Fossilised remains of mammals, leaves, trees and pollen spores have been found among the sediments there. The type of vegetation indicates that the climate in that region was more tropical at that time than it

PHOTO 12.3 Intertrappean sediments near Chilga, north of Lake Tana. The *white* layers are lake sediments, originally clay and silt, now altered to a siliceous rock called chert. The *darker* layers are rich in fossil vegetation on its way to becoming lignite, a form of impure coal. The total thickness of the sediments here is about 34 m. The basalt layers beneath and above them are not seen in this photo. *Hammer* for scale is just below centre (2014)

is today. It was warmer and wetter, giving rise to swamps and lush rainforests, with animals such as *Deinotherium* (an extinct type of elephant) roaming through them. The *Deinotherium* fossils found at Chilga are the oldest found anywhere in Africa.

Following this lull, volcanic activity recommenced over much of the region, but the erupted lavas tended to be more silica-rich, resulting in layers of ignimbrite and rhyolite rather than basalt. Sometimes these later lavas erupted through central vents, and the clogged interiors of these stand as plugs which punctuate the highland landscape (PHOTO 12.4).

Occasionally opals are found within the layers of the Trap Series volcanics, particularly the later, silica-rich flows. Opal is produced when water deposits dissolved silica, in the form of a gel, in cracks and cavities in a rock. The gel solidifies into trillions of minute spheres, packed together with water filling the spaces between them. It is the refraction of light by these tiny spheres that produces

PHOTO 12.4 "God's Finger" (Ye'Ab Idj), a volcanic plug north of Gonder. Many plugs like this, though few of them as spectacular, dot the Ethiopian landscape (2014)

the play of colours characteristic of opal. The conditions necessary for the formation of opal have to be very finely balanced, and this seldom happens which is why opals are rare and hence valuable. Ethiopian opal forms in a wide variety of colours ranging from pale blue to deep red and is very beautiful (PHOTO 12.5), and is highly prized on the world opal market.

The high plateau resulting from the lava flows and the uplift did not remain a plateau for long. As the land became higher, rivers simultaneously cut downwards through it. Although today quite extensive areas of the highlands do remain level to undulating (PHOTO 12.6), they are likely to plunge without warning into a deep

PHOTO 12.5   Opal from the Ethiopian highlands. The specimen is about 3 cm across. Photo courtesy of the South Australian Museum

PHOTO 12.6   Undulating, fertile land of the Western Highlands near Gonder (2008)

PHOTO 12.7 This photo, taken on a flight between Gonder and Addis Ababa, shows how rivers are dissecting what was once a plateau. The precise location of the photograph is not known (1971)

gorge. No-one driving the fairly level 200 km from Addis Ababa to the Blue Nile, for example, can fail to be taken by surprise at suddenly coming upon the lip of the precipitous Blue Nile Gorge. The situation is best appreciated from the air. PHOTO 12.7 was taken on a flight from Gonder to Addis Ababa and, although the location is uncertain, it shows clearly how rivers are eating their way into and across a gradually dwindling plateau.

12.2 The Shield Volcanoes and the Semien Mountains

Adding to the already considerable elevation of the highlands are a number of great shield volcanoes, successors to the Trap Series eruptions. Prominent among these are Mounts Mangestu, Choke, Guna, Guguftu and the Semien (Fig. 12.1). All these rise to elevations of over 4000 m and are formed mainly of basalt, with some flows of more silicic lavas and ignimbrite, particularly in their upper layers. The volcanoes vary quite widely in age. Indeed, it is hard to give "an age" to a volcano, as almost any volcano has periods of activity and quiescence over a long stretch of time. There may be a gap of millions of years between the youngest flows and the oldest ones, so the ages only give a very rough idea. Mt Guna, at 10.7 million years, appears to be the youngest of the shield volcanoes, Mt Mangestu is about 22.4 million years old, and the Semien massif is the oldest. In fact the oldest lavas of the Semien are dated at 30 million years, so there is little or no time gap between these and the latest eruptions of the Trap Series which underlie them.

The Semien Mountains are the highest in Ethiopia. They are customarily referred to in the plural as the original volcano has been eroded to numerous peaks. The reported elevation of the highest peak, Ras Dashen, ranges between 4543 and 4620 m depending upon which report one reads, but the exact height doesn't really matter as there is no doubt that the mountains are very high, majestic and spectacular (PHOTO 12.8). The main reason for their spectacular scenery is, rather ironically, the fact that much of the original volcano has been removed. The satellite

PHOTO 12.8 Crags, gorge and lava flows of the Semien Mountains east of Sankober (2007). Photo courtesy Emile Farhi

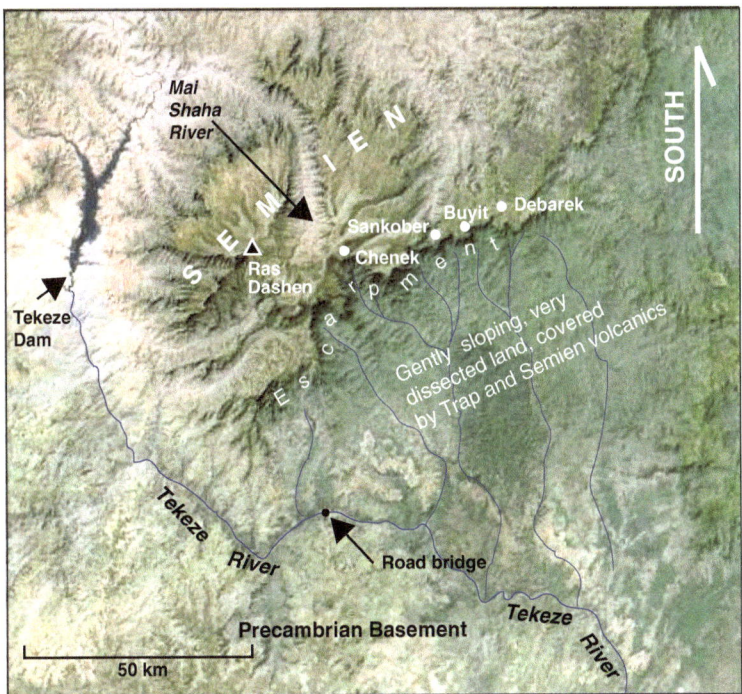

Fig. 12.2 Satellite image of the Semien massif and surrounding region. Note that the image is inverted, so that south is at the *top*. To the northwest (*lower right* in the image), half the originally circular volcano has been cut away by the tributaries of the Tekeze River, forming a great escarpment, and the remaining part of the volcano has almost been cut in half by the Mai Shaha River which probably follows a fault line. Image from Google Earth

image in Fig. 12.2 illustrates this. (Note that the image is oriented with south at the top.) Assuming that the original volcano was more or less circular, with a crater in the region of the present highest peaks, the entire northwestern segment of the volcano has gone. There is no evidence of a fault having been responsible for this—it is the work of erosion, almost entirely by water. The Tekeze river has carved its course to avoid the hard basalt flows of the volcano and its underlying Trap Series, cutting its gorge into the softer Mesozoic sandstone and Precambrian basement rocks to the north. This can be seen quite clearly on Fig. 11.1 in Chap. 11. Tributary streams flowing down the mountainside to join the Tekeze have gradually cut their

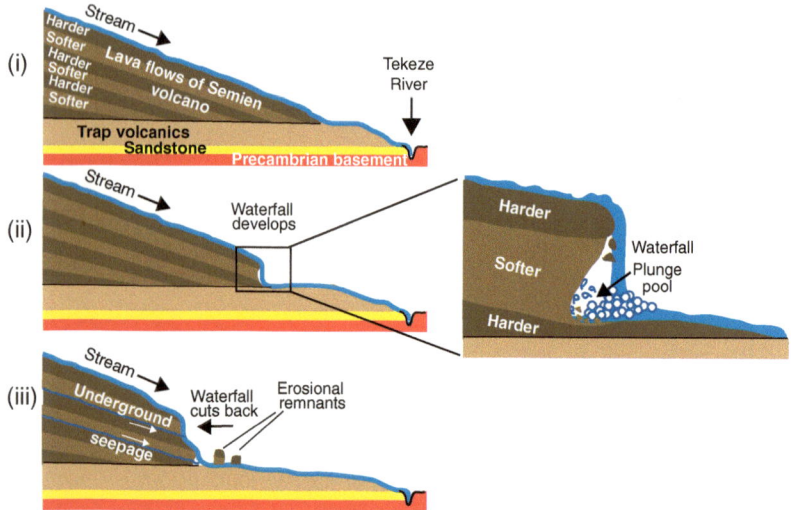

Fig. 12.3 Illustrating how headward erosion and spring sapping may have produced the Semien escarpment. (**i**) Tributary streams flow down the side of the Semien volcano, to join the Tekeze River. Although all of the lava layers are hard, some are slightly softer than the others. (**ii**) The stream cuts its way through a hard layer and encounters a softer one. The softer rock is more easily eroded and a waterfall is produced, a small one at first, but the force of the water erodes away the softer rock to produce a plunge pool, and splashing and turbulence wear back the softer layer to produce an overhang. Pieces of the hard layer fall down from the overhang, so that the waterfall gradually moves upstream. (**iii**) The same thing happens the next time the stream erodes through to a softer layer, so that a series of step waterfalls form. This eventually becomes a cliff as the lower waterfalls gradually move back to meet the upper ones. As the open ends of lava flows become exposed, water seeps between the layers and emerges in the form of springs, which accelerates the process. Note that the diagram is schematic only, and not to scale

way upstream into the mountain by a process called headward erosion, which is illustrated in Fig. 12.3. In this way they have created their own gorges with their own tributary streams, which in turn cut further back into the mountain. At the same time water seeps between the lava flows, emerging as springs at their exposed edges and further aiding erosion of the rock. This process is known as spring sapping. Together these processes have resulted in the scalloped escarpment of steep, stepped precipices and sheer cliffs which provide the breathtaking panoramas enjoyed by Semien trekkers (PHOTOS 12.9, 12.10 and 12.11).

PHOTO 12.9 The Semien escarpment, looking northeastward from a point between Buyit and Sankober. The escarpment plunges in steep steps for over 1500 m down to the lowlands. The spire in the centre of the picture is an erosional remnant (2009)

The remaining part of the mountain has been cut in two by the Mai Shaha River, which eventually flows into the Tekeze on the southeastern side of the mountain. Its very straight course through the flank of the mountain suggests that it follows a fault line. The Mai Shaha gorge presents a formidable obstacle to the traveller wishing to reach Ras Dashen summit. From Chenek camp on the western side of the gorge, the summit is less than 20 km away as the crow flies, and only about 500 m higher. So near and yet so far—for to reach it involves a descent of some 1200 m to the bottom of the gorge, followed by a gruelling ascent of more than 1500 m on the other side.

It has been estimated that the original Semien shield volcano, before erosion commenced, would have been about 110 km in diameter and perhaps 500 m higher than the summits we see today. Its crater, now completely eroded away, was probably slightly north of the present highest peak, Ras Dashen.

PHOTO 12.10 These cliffs near the top of the escarpment, near Chenek camp, look as though they have been sliced with a knife, resulting in a sheer drop of some 1000 m (2010)

Although little or no real snow falls on the Semien today, there is evidence that 20,000–30,000 years ago permanent snow covered its peaks above 4200 m, and that small glaciers were active between the highest summits. This was the time of the last of the great ice ages, when ice covered much of Europe and North America. The Semien glaciers have left evidence in the form of tillite (a jumbled mixture of ice-transported rock fragments in a clayey matrix, just like the Palaeozoic tillite which we met in Chap. 6), scratched boulders, and moraine ridges of boulders and fine sediment left behind as the glaciers retreated. The glaciers melted away about 15,000 years ago when the climate became both warmer and wetter. Although the glaciers are long gone, the high Semien can still be very cold. Water contained in crevices in the rocks often freezes during the night, causing the rocks to crack and eventually shatter into sharp fragments. The occasional gunshot-like sound of cracking rocks can be quite startling to campers!

PHOTO 12.11   Peaks of Imet Gogo on the Semien escarpment northwest of Chenek (2007)

12.3   The Southern Part of the Western Highlands and the Dividing Line

A careful look at Fig. 12.1 shows that the southern part of the Western Highlands appears rather different from the northern. South of the latitude of Addis Ababa there are no shield volcanoes and the terrain appears generally smoother, except towards its southern limit, southwards of Jima, where it is dissected by deep gorges as it grades into the plains of South Sudan and northern Kenya. There are also some differences in the rock types. The Trap Series volcanics become thinner, and generally lie directly on the Precambrian Basement. Only a very few outcrops of Mesozoic sedimentary rocks have been found in this southern region of the Western Highlands. Silica-rich rocks such as rhyolite and ignimbrite are more abundant than basalts. Although there are no shield volcanoes, there are numerous plugs and domes of trachyte and rhyolite.

Not only are the northern and southern parts of the Western Highlands geologically rather different, there is a distinct dividing line between them. A conspicuous linear feature, which is marked on Fig. 12.1 by a white dotted line, extends from the region of Addis Ababa to the border with South Sudan. It is known by the rather formidable name of the Yerer-Tulu Welel Volcanotectonic Lineament, which translates as a "linear region of volcanoes and faults that runs between Yerer volcano (near Addis Ababa), to Tulu Welel volcano in far western Ethiopia". It is one of a number of similar features that run from east to west across Ethiopia, and it is thought that these all occur along lines of weakness that originated during the collision events of the Precambrian era over 500 million years ago. Figure 12.4 shows a digital elevation model of the Yerer-Tulu Welel Volcanotectonic Lineament, and it is worth taking a close look at this structure as it contains some interesting and attractive features.

The lineament is not a simple narrow line, but a broad belt of faults and volcanoes, about 80 km wide and 700 km long. Part of its northern edge is marked by an escarpment, beginning from the Entoto Mountain which forms a dramatic backdrop to Addis Ababa, and extending as an almost continuous feature to beyond Ambo. It can be seen clearly on the right-hand side of the road from Addis Ababa to Ambo, and is sometimes called the Ambo Fault. The actual fault is very likely well south of the escarpment, passing through the town of Ambo itself and

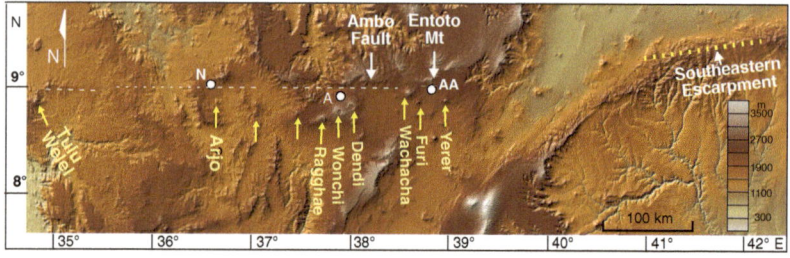

Fig. 12.4 Digital elevation model of the Yerer-Tulu Welel Volcanotectonic Lineament, showing the Ambo Fault and the volcanoes associated with the Lineament. *AA* stands for Addis Ababa, *A* for Ambo and *N* for Nekemte (*Yellow arrows* indicate volcanoes. Two are not named). DEM from GeoMapApp

being responsible for the springs which feed the swimming pool and provide the famous Ambo mineral water. Addis Ababa's Filweha springs, one of the features that influenced the choice of this location for Ethiopia's capital city, are also located on this fault. The original fault scarp has been eroded and gradually moved backward. The land on the northern side of the Ambo Fault has moved up about 800 m relative to that on the southern side (or vice versa), in places exposing the Mesozoic sandstones which underlie the Trap Series volcanics. These can be seen in a quarry near to Ambo. North of the fault scarp the land tilts slightly northward, a point which will become significant when we look at the course of the Blue Nile River in the next chapter.

Close to Ambo is a beautiful little volcano called Wonchi, one of the several volcanoes associated with the lineament and shown on Fig. 12.4. It has a deep crater lake, and a crater rim formed almost entirely of white volcanic ash, less than a million years old (PHOTOS 12.12 and 12.13). Hotsprings around the lake-shore confirm that volcanic activity has been very recent. Wonchi is actually part of a larger volcanic complex called Regghae Badda. A small and equally pretty double crater called Dendi, about 10 km east-north east of Wonchi, is also part of this complex. The volcano Tulu Welel, which we met in Chap. 10, is the westernmost manifestation of the lineament. The lineament may continue further; if so it is buried beneath the sands of South Sudan.

It is interesting to note that the Yerer-Tulu Welel Volcanotectonic Lineament is almost, though not quite, in line with the southern escarpment of Afar. Whether this is a coincidence, or significant, remains an intriguing matter for conjecture.

PHOTO 12.12 Wonchi, a small volcano with a crater lake, near Ambo. There are actually three craters nested together and the crater rim is formed of white ash. Wonchi is part of the Yerer-Tulu Welel Volcanotectonic Lineament described in the text (2013)

PHOTO 12.13 Layers of white ash forming the rim of Wonchi volcano (2013)

In the next chapter we will move to the northern part of this highland region, to visit Ethiopia's largest lake and her most famous river, the Blue Nile.

Lake Tana and the Blue Nile

The Blue Nile, or Abay, is Ethiopia's best-known and most revered river. It rises in the Western Highlands and turns back on itself in a great loop before descending to the Sudan plains and meeting with the While Nile at Khartoum. In Chap. 9 we saw how the deep gorge which it has cut along its way has revealed much of Ethiopia's geological history. In this chapter we will look at the river itself, where it comes from and how it has carved not only a great gorge, but what appears to be an oddly circuitous course through the highlands as illustrated in Fig. 13.1. We will start in its headwaters: the basin of Lake Tana.

13.1 Lake Tana

Lake Tana is by far the largest lake in Ethiopia. It covers an area of 3156 km^2 and contains half of the fresh water in the country. It is about 84 km long and 66 km wide but is surprisingly shallow, having an average depth of only 9 m and a maximum depth of 14 m. These depths are however rather misleading, since at least 100 m of sediment covers the bedrock which forms the original lake floor. Before these sediments were carried in by rivers draining the surrounding highlands, the lake would have been considerably deeper. Also, evidence from ancient shorelines indicates that just under 15,000 years ago the lake was twice as large, in area, as it is today, and its surface level about 75 m higher. Much of the present lake is still bordered by swamps and wetlands, remnants of its larger ancestor. Why then is such a big lake perched high up in the Western Highlands?

© Springer International Publishing Switzerland 2016
F.M. Williams, *Understanding Ethiopia*, GeoGuide,
DOI 10.1007/978-3-319-02180-5_13

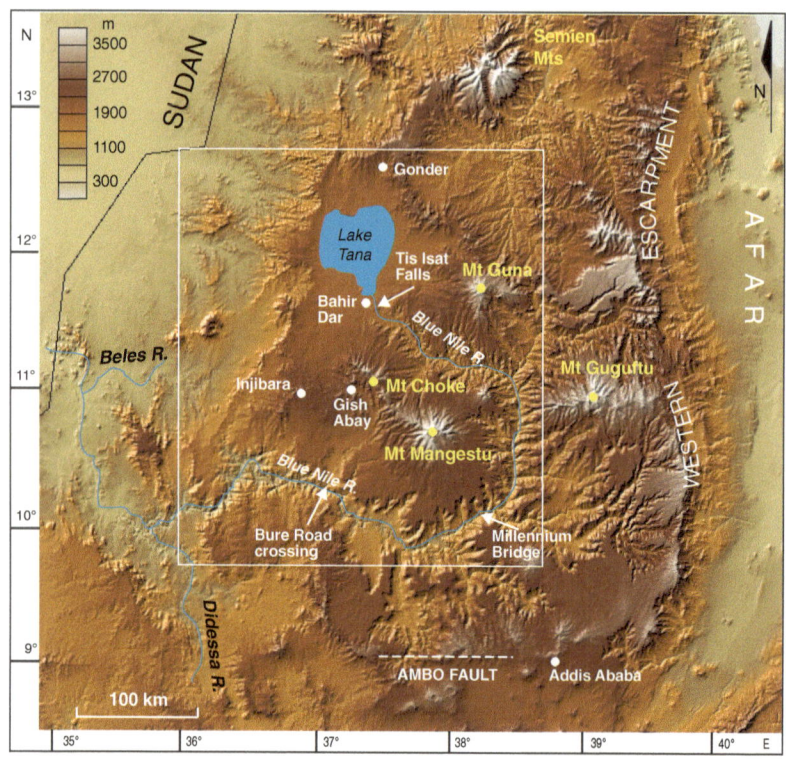

Fig. 13.1 Digital elevation model showing the region around Lake Tana, and the course of the Blue Nile from Lake Tana to the Sudan border. The *white rectangle* indicates the area covered by the geological map in Fig. 13.2. DEM from GeoMapApp

Figure 13.2 shows a geological map of the region around Lake Tana. To the east and south of the lake are the big shield volcanoes of Guna and Choke (pronounced Chok'uay)[1]; and to the southwest is a region of fresh-looking volcanic rocks: lavas, cinder cones and plugs. To the west is a rather narrow ridge beyond

---

[1]There is some confusion with the names of these two mountains, which I have been unable to resolve. On some maps the mountain I have marked as "Mangestu" is named as Choke. The mountain I have marked as Choke is sometimes not named at all, or marked as Choke or Amedawit. On yet another map both mountains are grouped together as "Choke Mountains". Questioning local people has not helped to resolve the issue. The usage I have adopted is that of the Geological Map of Ethiopia (1st ed., 1973) (the 2nd edition does not name either of the mountains) and of the Geological Map of Ethiopia and Somalia (Merla et al. 1973).

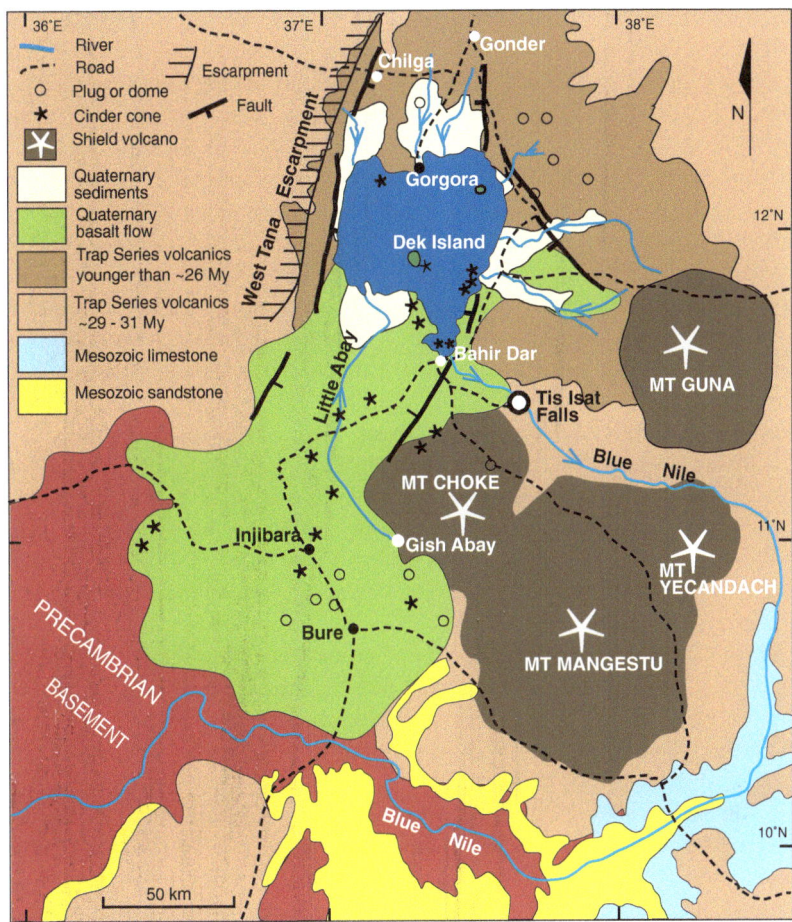

Fig. 13.2 Geological map of the Lake Tana region. Modified from Geological map of Ethiopia 1:2,000,000 2nd Ed. (1996), Geological Survey of Ethiopia

which a steep scarp, the West Tana Escarpment, faces toward the lowlands which slope down to the Sudan plains.

There are numerous faults in the region, but only the main ones are shown on the map. Lake Tana lies at the intersection of three fault systems—one running NW-SE, one nearly N-S and the third NNE-SSW. Not all the faults are easy to distinguish on the ground, or even on satellite images, as later lava flows have concealed them, but it is thought that they outline three rifted basins. If you look back to Fig. 6.2 you will see that the NW-SE one is inherited from Mesozoic times, when Gondwana was beginning to break apart. There is no evidence that the other two are as old as this and it seems more likely that they formed later, perhaps when Arabia began moving away from Ethiopia and the whole region was pulled and stretched. This also caused the NW-SE faults to move once more, and at some stage the land at the intersection of the three rifts, weakened by the faulting, subsided to form a rectangular-shaped basin.

Rivers flowed into the basin from the flanks of Mt Guna and Mt Choke, from the high land around Gonder and from the ridge bounding the basin to the west. As yet, however, there was nothing to stop them from flowing out of its southern end. Two events led to the birth of Lake Tana as we now know it. Firstly, as the region continued to be pulled and stretched, the basin subsided a little more between its bounding faults, and sagged in the centre. Then a voluminous eruption of basalt lava blocked any remaining southern outlets. The lava, shown in green on Fig. 13.2, originated from numerous centres now seen as the plugs, craters and cinder cones which dot the countryside between Bure and Lake Tana, and sometimes lava tunnels have formed where liquid lava kept on flowing beneath a solidified crust (PHOTO 13.1).

Although to my knowledge this lava field has not been dated (though numerous guesses have been made), its fresh appearance suggests that it is less than 2 million years old, and on published geological maps it is shown as Quaternary in age. It extends northwards across much of the basin area which is now occupied by the lake (PHOTO 13.2), and most of the thirty seven islands which dot the lake are formed of its blocky fragments, or are cinder cones that protrude above the lake surface (PHOTO 13.3). These islands are very important as over half of them provide secure and secluded locations for ancient and beautiful churches and monasteries (PHOTO 13.4).

PHOTO 13.1 Zen Akwashita "cave", actually a lava tunnel, near Injibara south of Lake Tana. Liquid lava continued flowing after the top of the flow solidified, and when the source of the lava ceased eruption the liquid drained out leaving a hollow tube. An underground stream now flows through the tunnel whose length is not known but is said by local people to be "very long" (2014)

Since Lake Tana first formed, rivers have been spreading their sediment loads across its floor. Cores taken by drilling into these sediments have revealed an interesting story, both of the lake itself and of the changing climates that affected it. The deepest sediments reached by coring, at a depth of 92 m, have been dated at 250,000 years, so the lake must be at least as old as this. Dating of the lava which barred its southern outlets would be very useful in helping to establish the age of the lake. Detailed study of the sediment cores shows that the lake was full and even

PHOTO 13.2 Boulders of Quaternary basalt on the shore of Lake Tana at Bahir Dar (2014)

overflowing from about 130,000 to 95,000 years ago, but 16,400 years ago it dried out completely indicating that, as elsewhere in Ethiopia, this was a time of very dry climate. After this the lake filled again, and about 15,000 years ago overflowed once more, at the lowest point on its rim—the outlet to the Blue Nile. Since 1996 the lake level has been regulated by the construction of a control weir at the Blue Nile outlet.

It must be borne in mind, however, that these are only the latest events. What happened beforehand, how many times the lake dried out, and how many times it may have overflowed and ceased to overflow into the Blue Nile, remain matters for conjecture until deeper cores can be obtained to penetrate further into its history.

PHOTO 13.3 Dek Island, the largest of Lake Tana's islands, built of blocks of scoriaceous basalt. It is a flattish "hump" in the field of Quaternary lava that extends across the lake basin (2014)

Geologically, Lake Tana is in rather a precarious position. Only a dozen or so kilometres from its western shore is the West Tana Escarpment, which essentially marks the limit of the Western Highlands. This shows clearly on Fig. 13.1. This escarpment, like that which bounds the northwestern side of the Semien Mountains and which branches from it, is the result of headward erosion, in this case by streams flowing toward the Sudan lowlands. Such an escarpment is known as a retreat scarp, because it is doing just that—retreating. Erosion is slowly wearing it back, and when it reaches the lake in a few million years' time it may spell the end of Lake Tana!

PHOTO 13.4   The 18th century monastery of Narga Sellassie on Dek Island. This beautiful church is built of ignimbrite which had to be ferried to the island from the mainland north of Gorgora in papyrus boats, as the scoriaceous basalt of the island was not strong enough for its construction. Most of the monasteries on Lake Tana's islands are much older than this one (2014)

13.2   The Source of the Blue Nile

During the 17th and 18th centuries a number of intrepid explorers, including the ebullient Scotsman James Bruce, claimed to have "discovered" the source of the Blue Nile. This was an instance of European arrogance, since the Ethiopians knew it perfectly well all along: a group of springs near the village of Gish Abay, about 75 km south of Lake Tana. These springs feed a small river known as the Little Abay, one of a number of such rivers which flow into the lake. This has raised queries with geographers—should not the headwaters of those other rivers have an equal claim to be the source? Or since so many streams flow into the lake (about 60 according to one count), whereas the Blue Nile is the only one that flows out of it—shouldn't Lake Tana itself be considered as the source?

As with all rivers, the water of the Blue Nile comes from numerous origins, increasingly so as it flows downstream. In fact, by the time it joins the White Nile at Khartoum only 8 % of its total flow has come from Lake Tana—and of that only a small fraction has come from the Little Abay. The vast bulk of its water has come from tributaries that joined it on the way, some of which such as the Didessa and Beles are major rivers in their own right.

In the end such technicalities don't really matter. To Ethiopians the springs at Gish Abay are the source of the Blue Nile, ordained by heaven, and pilgrims come from all over the country to receive healing from their sacred waters. The hillslope at Gish Abay is a beautiful spot, and a fitting place for such a great river to begin its long journey (PHOTO 13.5).

PHOTO 13.5 The Little Abay emerging from springs at Gish Abay. The springs themselves are concealed in stone and tin shelters which pilgrims enter to sample the holy waters (2008). Photo courtesy Emile Farhi

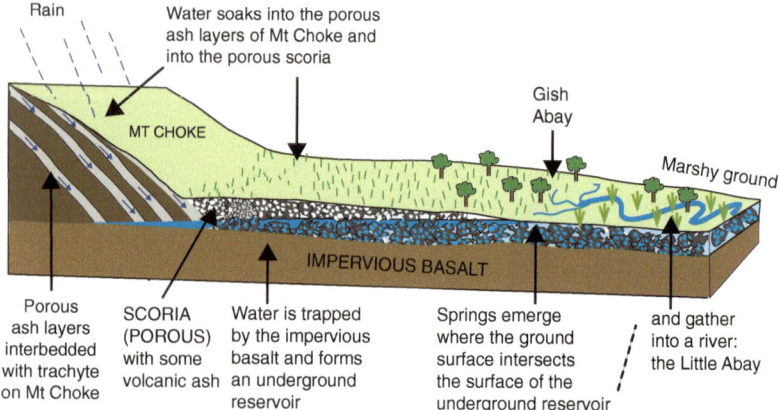

Fig. 13.3 Sketch showing the formation of the springs at Gish Abay

More prosaically, the springs at Gish Abay are produced by water which has seeped out around the foot of Mt Choke and from the lava that covers the area south of the lake. This is illustrated in Fig. 13.3. Much of the lava is in the form of blocky basalt boulders containing many cavities, or vesicles. Such a texture in basalt is known as scoriaceous, and because of the vesicles the rock is very porous. Water is able to seep into it, but is stopped from going deeper by the layer of impervious basalt below and bubbles out to the surface as springs.

13.3   The Blue Nile and the Great Loop

"And the name of the second river is Gihon: the same is it that compasseth the whole land of Ethiopia."[2] This passage from the book of Genesis describes one of the four rivers that branched from the Garden of Eden. Although, as we saw in Chap. 12, the Ethiopia of the Bible does not correspond geographically to today's Ethiopia, it is quite possible to see why Ethiopians have adopted the river Gihon (usually spelled Ghion in Ethiopia) as their own and identified it with the Blue Nile. Though obviously not encompassing the whole of present day Ethiopia, the

---

[2]This is the King James version. In most modern translations the name "Ethiopia" is replaced by "Cush".

Blue Nile does make a great loop around a large part of her Western Highlands and its tributaries extend over an even larger area as can be seen on Fig. 13.1.

At first sight the course of the Blue Nile, in making this great loop, appears oddly convoluted. First it flows southeastwards towards Afar but turns south before it reaches the Western Escarpment. It then curves around toward the northwest, almost reversing its original direction as it circles around the Choke and Mangestu mountains. Did it once flow to the Red Sea, and then reverse its course as the highlands were uplifted and the high ridge bordering the Western Escarpment formed? Maybe, but there is no evidence for this. It is clear from Fig. 13.1 that the present-day Blue Nile is simply following the topography, its course confined firstly by the shield volcanoes of Guna, Mangestu and Choke, then by the high shoulder of the Western Escarpment and finally by the northward tilt resulting from the Ambo Fault. But how long has it been following this course, and how did it carve such a deep and wide gorge? At the Millennium Bridge crossing, the top of the gorge spans a distance of about 20 km whereas the river at the bottom, 1500 m below, is only about 100 m wide even when full. Even the fact that the river was certainly larger and more powerful during wetter periods in the past cannot account for the size of the gorge.

The answer to both these questions almost certainly lies in headward erosion, the process by which a river may cut its way upstream as we saw in Chap. 12, acting in concert with the geological events that were taking place as the river developed. The Afro-Arabian Dome began to rise in the mid-Eocene, about 40 million years ago. The Blue Nile may have started as quite a small river, flowing off the western slope of the rising dome and at the same cutting its valley into it. Rivers inevitably cut downwards, attempting to reach their base level which ultimately is that of the sea. As the dome continued to rise and the eruption of the Trap Series volcanics added to its height, the river simultaneously cut downwards and headwards, gradually carving its gorge upstream. By about 25 million years ago the gorge had reached the present day Bure road crossing and it would have reached the present position of the Millennium Bridge between 20 and 15 million years ago. Tributaries joined the river, contributing to its flow and creating their own gorges as they too cut into the rising land.

As the gorges deepened, they also widened as material from the steep and unstable sides broke away to be carried off downstream. Although this process took place only very slowly, 25 million years has been time for a lot of material to be eroded away. It has been estimated that the Blue Nile and its tributaries have removed altogether about 90,000 km^3 of rock from the Ethiopian highlands! This process is still occurring today, and the Blue Nile Gorge is one of the most landslide-prone regions in Ethiopia.

PHOTO 13.6   The Tis Isat ("smoke fire") Falls, where the Blue Nile tumbles over a barrier of Quaternary basalt and is transformed from a broad, gently meandering river near the *top* of the picture to a roaring torrent confined in a narrow gorge at the *bottom left* (2008). Photo courtesy Emile Farhi

At the present time the gorge has almost, but not quite, reached Lake Tana. About 25 km short of the lake it has encountered an obstacle in the form of a flow of hard Quaternary basalt, which has prevented or at least delayed it from cutting back further. It is this hard basalt layer which is responsible for the Tis Isat Falls, one of Ethiopia's most popular tourist attractions. Above the falls the Blue Nile is a broad and gently meandering river, below them it is confined as a rushing torrent in its narrow gorge (PHOTO 13.6). Unfortunately the falls are less magnificent than they used to be, as water has been diverted from them to operate a hydroelectric power plant.

The Blue Nile provides a lifeline for Sudan and Egypt since it, together with the Tekeze River, not only contributes 70 % of the water of the main Nile, but carries down from the Ethiopian highlands 95 % of the mineral-rich sediment that forms the fertile soils along the Nile Valley. Each year, during Ethiopia's rainy season, these soils are replenished. Failure of the Blue Nile floods is a disaster for these countries downstream. 4200 years ago a severe drought caused the White Nile to cease flowing and the Blue Nile floods to fail, leading to the collapse of the Old Kingdom of Egypt. Careful thought must therefore be given to the consequences of building dams or initiating other projects that might disrupt the Blue Nile's flow.

In the next chapter we will move eastward from Lake Tana to our final locality on the Western Highlands: the town of Lalibela and the unique geology which enabled construction of the rock churches for which it is renowned.

The Geology and Churches of Lalibela

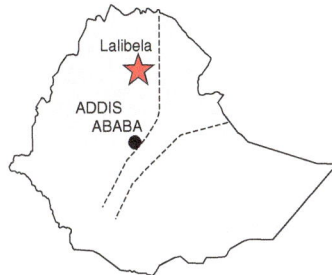

According to legend Lalibela, who became king of Ethiopia in the 12th century, was requested in a heavenly vision to build a new Jerusalem of rock at Roha in the Ethiopian highlands. He obeyed, resulting in the wonderful rock-hewn churches that we see at Lalibela today (PHOTO 14.1).

The churches were not, however, built in the usual sense of the word, but excavated from the solid rock. Figure 14.1 illustrates the concept. Although all eleven[1] of the churches are different in design, the principle was the same for each. Be it a free-standing church, such as Bet Giorgis or Bet Medhane Alem, or a semi-monolithic one whose roof or a wall remained connected to the bedrock, such as Bet Libanos, it was constructed by first excavating a deep trench to produce an isolated or semi-isolated block. The interior of this block was then hollowed out, leaving rock in place to form pillars, arches and interior walls, all of perfect

---

[1]The number of churches is sometimes quoted as 13 if the joined churches of Bet Mikael and Bet Golgotha are counted separately, and the Selassie Chapel within Bet Golgotha is included in the count.

© Springer International Publishing Switzerland 2016
F.M. Williams, *Understanding Ethiopia*, GeoGuide,
DOI 10.1007/978-3-319-02180-5_14

PHOTO 14.1 The rock hewn church of Bet Giorgis, sculpted in the shape of a cross. A figure standing close to the *right side* of the trench indicates the scale. This church is reputedly the last one to be constructed at Lalibela, to appease an angry St George to whom King Lalibela had omitted to dedicate any of the previous ten churches (2014)

symmetry and often very elegant design. It was an incredible undertaking, particularly as the churches are not small. Bet Medhane Alem, for example, measures 33.5 m by 23.5 m and is 10 m high, and is reputed to be the largest rock-hewn church in the world. Just who built them, how long it took, and who orchestrated the mammoth task is a matter for historians, but the method is clear from the markings on the walls of the churches—they were carved out by use of simple hammer and chisel.

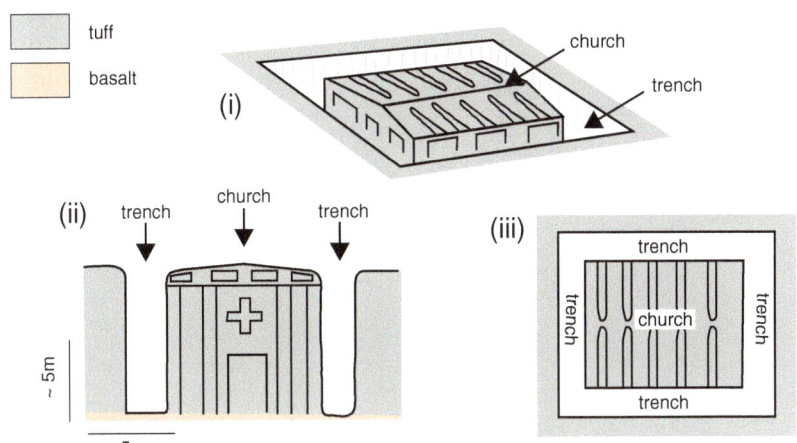

Fig. 14.1 Generalised structure of a rock-hewn church in (**i**) three dimensional view, (**ii**) cross section view and (**iii**) plan view. A block of rock is isolated by excavating a trench around it. The inside of the block is then hewn out, leaving rock in place for pillars, arches and other interior decorations. Windows are carved into the walls, and symbolic structures such as window bosses, beams and outlines of stonework are carved into the walls. Some churches are not entirely isolated like this one is, but may still have a wall or the roof connected to the surrounding rock. Note that the scale only gives a general idea, as individual churches vary considerably in size

Whoever inspired Lalibela's vision, be it God or angels, it seems that they knew some geology for they specified one of the few places in that region where it would be possible to construct such churches. The geological sketch maps in Figs. 14.2 and 14.3 show why. The town of Lalibela is surrounded by, and built upon, Trap Series basalt, a very hard rock which would be almost impossible to excavate and sculpt. However, occupying a hollow in the basalt is a small area, only about half a kilometre squared, of orange-red tuff (PHOTO 14.2). Tuff is a material formed of volcanic fragments, probably the result of a brief but violent episode of fire-fountaining during the otherwise smooth eruption of the basalt flows. It consists of fragments of basalt, cindery fragments known as scoria, glassy shards and droplets, and mineral grains (PHOTOS 14.3 and 14.4). Even when compressed, tuff is much softer than basalt and it is this that enabled the churches to be carved.

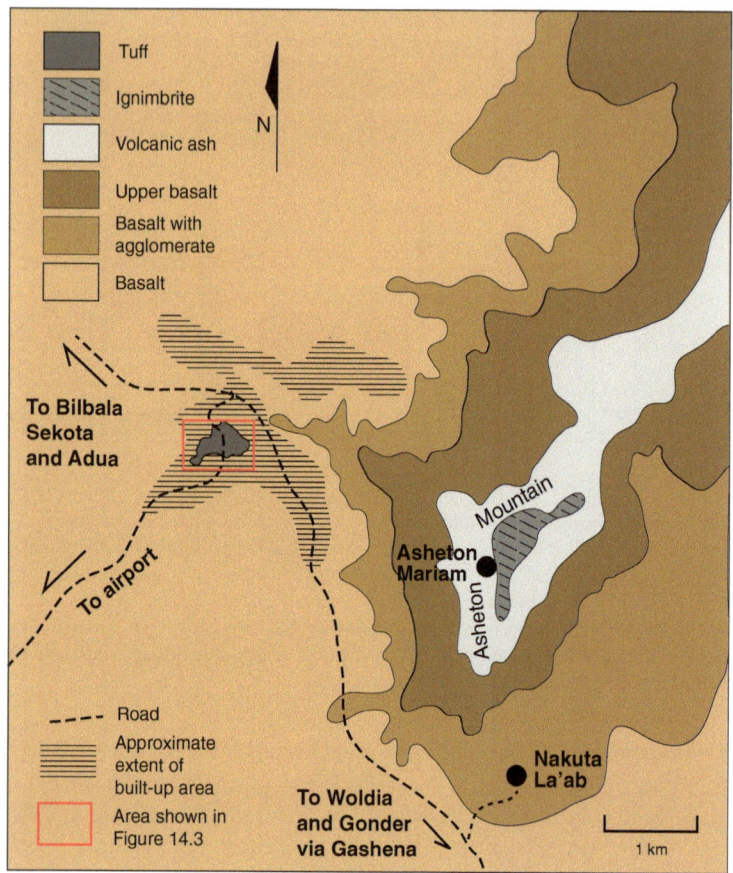

Fig. 14.2 Geological sketch map of the Lalibela region showing the small area of tuff in which the church complexes are located. The locations of two outlying churches, Asheton Maryam and Nakuta La'ab are also shown. Other outlying churches mentioned in the text are beyond the range of this map. Modified from Asrat and Ayallew (2011)

That is not to say that the task would have been easy. The tuff is very firmly consolidated, otherwise the churches could not have lasted until today, but it is vastly more amenable to chipping by chisel and hammer than basalt. The deposit is not very thick, however, and in some of the trenches which surround the churches it can be seen that the underlying basalt was reached, limiting the excavation (PHOTO 14.5).

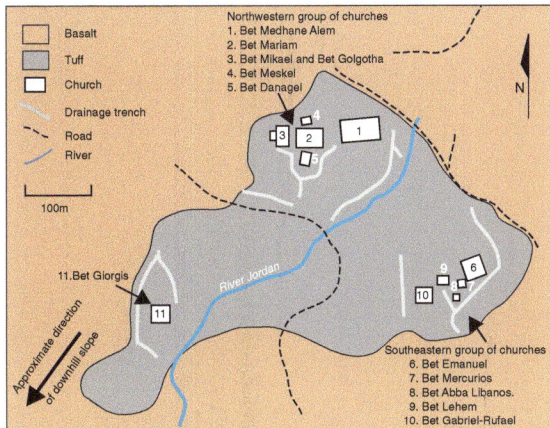

Fig. 14.3 Sketch map showing the arrangement of the rock-hewn churches at Lalibela. The churches are clustered into three groups as shown. Within each group, excepting Bet Giorgis which is isolated, the individual churches are linked by an intricate network of trenches and tunnels. Tuff boundary from Asrat et al. (2008)

PHOTO 14.2 View of Lalibela from Asheton Mountain, showing the area of red tuff into which the churches are excavated. This photo was taken in 1969. It is difficult now to obtain an unimpeded view, as much of the area is concealed by expansion of the town (1969)

PHOTO 14.3 Close-up of tuff in the wall of Bet Giorgis trench. The *light coloured fragment* on which the one *birr coin* is resting is a rock fragment; the *dark patch* near the *top right* of the photo is a piece of scoria (2014)

The original builders of the churches were also good engineers. The churches are located on the downslope of a hill, so that natural surface runoff would flow across the church complexes. Aware of this, the constructors put in place a carefully planned system of trenches to drain water away from the churches, enabling their preservation for over 800 years. Sadly, the churches have deteriorated more during the past few decades than they have since the time of their construction. Some patchy repairs were carried out by the Italians during and after their occupation of Ethiopia, but in more recent years the population of the town, its accessibility, the amount of traffic and the number of visitors to it have increased hugely, to the detriment of the churches. Many of the old drainage trenches have

PHOTO 14.4 A volcanic bomb (a large blob of lava that has solidified in flight) from the Bet Giorgis trench. A one *birr coin* placed on the bomb, just above the centre of the photo, indicates the scale (2014)

become clogged by debris and vegetation. Vibrations from traffic and from the activities of people have caused movement along joints and fractures in the church walls (PHOTO 14.6). Pressure from the growing population has altered the water-table conditions, and changes in land-use patterns have increased the seepage of groundwater into the church complexes. Ironically, the properties of the tuff itself contribute to the problem. It is more porous than the basalt which underlies it so that water tends to concentrate along the tuff-basalt boundary, causing damage to the base of the church walls (PHOTO 14.7). The churches of the southeastern

PHOTO 14.5 The contact between the *pinkish-brown* tuff and *grey* basalt at the base of Bet Medhane Alem. The basalt limited the depth to which the churches could be excavated (2014)

group (Bet Emmanuel, Bet Mercurios, Bet Lehem, and Bet Gabriel-Raphael) are more damaged than those of the northwestern group because they are located close to the edge of the tuff deposit where it is less consolidated as well as thinner, and therefore more prone to water percolation.

In an attempt to remedy the situation, large (and in my opinion ugly) shelters have been built over the churches, but this has failed to recognise that rainwater damage from above is only part of the problem. The attack is also coming from below. The ancient builders with their system of trenches showed insight that has been lacking in today's rush for growth and development. In a fresh attempt to address the problem, more drastic measures are currently being taken by relocating

PHOTO 14.6 A crack in the wall of Bet Raphael. The marker was placed across the crack in 1968 to monitor the rate of widening. Between then and 2014 it has widened by about 2 mm (2014)

the population away from the church complexes, and building a new road to divert heavy traffic.

Other churches in the Lalibela region are also located and/or constructed according to geology (Fig. 14.2). Asheton Mariam, for example, a small monastic church situated on Asheton Mountain overlooking the town, is carved into soft white volcanic ash (PHOTO 14.8). The ash forms a conspicuous layer running along the mountainside, and contains many fragments of fossil wood (PHOTO 14.9), remnants of trees buried during its eruption. The monastery of Nakuta La'ab, on the south side of Asheton Mountain is very different. It is built within a natural cave between layers of basalt and basalt agglomerate, a blocky

PHOTO 14.7 Damage by water seepage to the base of Bet Emmanuel. The cover erected over the top of the church, of which a portion can be seen, has not addressed this problem. The Italians attempted to repair the damage with cement, which has now been removed (2014)

PHOTO 14.8 The entrance to the monastic church of Asheton Mariam. The church is excavated into soft white volcanic ash. The church entrance is at the top of the steps above the head of the white-robed figure. The other small caves are cells for monks to rest and pray in (2008)

type of basalt. The cave has been extended a little at the rear, by chipping out the rock, and the church itself is built of bricks crafted from tuff (PHOTO 14.10). A stream flows above the cave and spills over as a waterfall in front of it, making it distinctly damp. Water continually drips through the cave roof into the church itself where bowls, carved from basalt, have been strategically placed to collect it (PHOTO 14.11). The supply never runs out, and it is used as holy water by the priests.

Other outlying churches are too scattered to show on Fig. 14.2. Some, such as the group of churches near Bilbala about 20 km[2] northwest of Lalibela, and Ganeta Mariam about 9 km to the south-southwest, are similar in design to the Lalibela

---

[2]Note that all distances given in this chapter are "as the crow flies". The road distances are considerably longer.

PHOTO 14.9 Large piece of fossil wood, about 20 cm in diameter, in white volcanic ash on the way to Asheton Mariam church (2008)

churches and have, like them, exploited patches of tuff within the surrounding basalt. East of Bilbala, the church of Imrahana Christos is built into a natural cave within basalt, taking advantage of the thick, impervious rock for protection of the building (PHOTO 14.12). The church itself is built of alternating layers of timber, and stone blocks (probably tuff) plastered with gypsum. The outlines of basalt columns are spectacularly exposed in the roof of the cave. Unlike Nakuta La'ab, water does not percolate through the thick roof, and so the church's wooden beams, window frames, and beautiful interior decorations have been perfectly preserved (PHOTO 14.13).

PHOTO 14.10 The monastery of Nakuta La'ab, built of bricks of tuff within a basalt cave. Water drips through the roof into the church and is collected in basins crafted from basalt (2014)

Along a cliff-side near the town of Checheho, about 60 km southwest of Lalibela, is a row of five adjacent small churches carved into light grey volcanic ash (PHOTO 14.14). Although neither large nor elaborately decorated, they are particularly interesting because they were excavated in recent times, between about 2005 and 2014, by a priest called Abba Dafar in response, like Lalibela, to a holy vision. According to local information the task took him and two assistants 5 years to complete, using only the simple type of tools that would have been available to the Lalibela constructors. This could, with a great deal of extrapolation since the

PHOTO 14.11 Basalt basins collect water that drips through the roof of Nakuta La'ab church (2014)

rock is far softer, the design far simpler and the scale much smaller, throw a glimmer of light upon how long it might have taken to construct the Lalibela churches.

There is no room here to describe all the churches around Lalibela, and I cannot claim to have visited more than a few of them. If you have a chance to visit other outlying churches look carefully at their construction, their state of preservation, and their setting, and consider how much of this would have been determined by geology.

PHOTO 14.12 The church of Imrahana Christos, built of timber beams (*brown*) and blocks of tuff plastered with white gypsum, within a basalt cave. Note the columnar jointing in the roof of the cave, at *top left* of the photo (2004)

PHOTO 14.13   The thick, impervious roof of the cave preserves the beautiful carvings and paintings within Imrahana Christos church from damage. This view is taken looking upwards, past one of the ceiling arches (2007)

PHOTO 14.14 A group of five churches near Checheho has been excavated in recent times, in a layer of grey volcanic ash. The churches are all quite small, and have some simple interior decorations such as pillars. From *left to right* their names are Gabriel, Maryam, Bel Egziabher, Raguel and Raphael (2014)

The Southeastern Highlands and the Ogaden

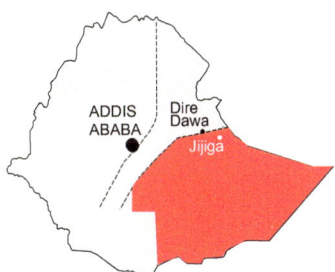

This chapter covers the region to the southeast of the Main Ethiopian Rift and Afar. It looks a big area to cover in a single chapter—about one third that of all Ethiopia—but a great deal of it is difficult to access and thus cannot be described in any detail. However, the parts which can be visited offer a variety of geological features and scenery which are not found elsewhere in Ethiopia. The people too are distinctive. Predominant over much of the region are the Somali people, their vibrant and colourful dress in contrast to the gentle white fabrics favoured by the inhabitants of the Western Highlands, and in the mountains bordering the Rift Valley are the Oromo, known in this region for their brilliant horsemanship.

It is clear from any topographic map or image of Ethiopia, such as that shown in Fig. 15.1, that the Southeastern Highlands[1] are much less extensive that the Western ones. In fact they constitute little more than a narrow ridge bordering the

---

[1]Because these highlands are, increasingly towards the east, inhabited by Somali people they are sometimes referred to as the Somali Plateau. I prefer the more geographically descriptive term of Southeastern Highlands.

© Springer International Publishing Switzerland 2016
F.M. Williams, *Understanding Ethiopia*, GeoGuide,
DOI 10.1007/978-3-319-02180-5_15

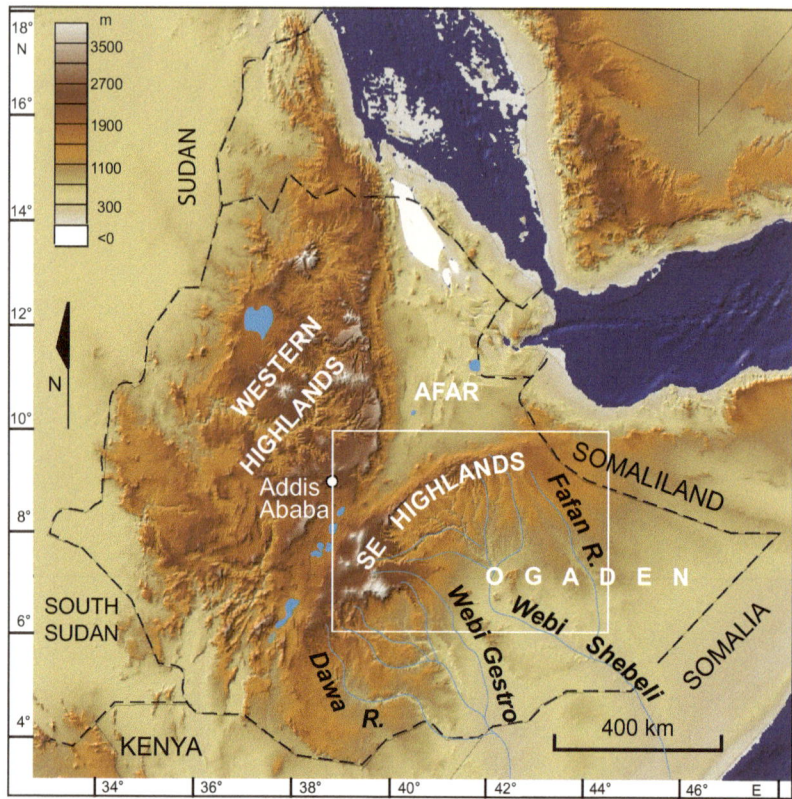

Fig. 15.1 Digital elevation model of Ethiopia showing the location of the Southeastern Highlands and the Ogaden. The *white rectangle* indicates the area shown in more detail in Fig. 15.2. DEM from GeoMapApp

southeastern margin of the Main Ethiopian Rift and the southern margin of Afar, shown in more detail in Fig. 15.2. The geological map in Fig. 15.3 shows that the volcanic rocks of the Trap Series, which are so ubiquitous over the Western Highlands, are more or less confined to this narrow ridge. This is because we are approaching the edge of the Afro-Arabian Dome and the thick pile of lava flows that cover it. Southeast of this ridge the original lava flows were much thinner and have since been almost entirely eroded away. The ridge itself is due both to the thickness of its lava cover and to uptilt of the escarpment rim, a phenomenon that will be explained in Chap. 20.

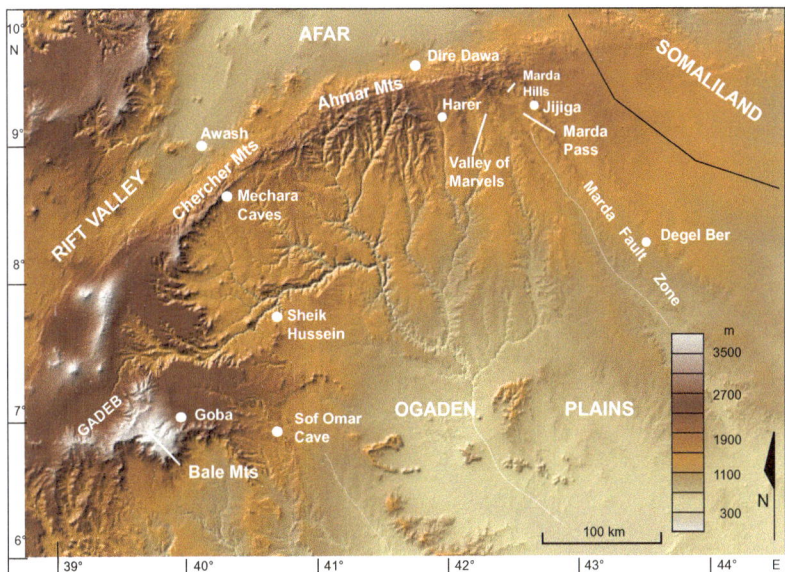

Fig. 15.2 Digital elevation model of the Southeastern Highlands and the northwestern part of the Ogaden. DEM from GeoMapApp

The predominant colour on the geological map is blue. This represents the Mesozoic sedimentary rocks, mostly limestones, that underlie the Ogaden, a region of semi-desert that extends from south of the highland ridge into Somalia. Along the ridge itself these sedimentary rocks are covered by Trap Series volcanic rocks, coloured light brown. The section of the ridge that borders the northernmost part of the Rift Valley constitutes the Chercher Mountain range, and that bordering the southern margin of Afar the Ahmar Mountains. The two ranges are separated by a slight dip, which provides convenient passage for the main road from Awash to Dire Dawa and Harer. Although they lack the precipitous canyons of the Western Highlands, these ranges have been eroded to an attractive landscape of mountain peaks, fertile plains and wooded valleys.

The Ahmar and Chercher Mountains form a watershed between the ephemeral streams which flow northward into Afar, and the headwaters of the Webi[2] Shebeli which flows southeastward toward the coast of Somalia. Headward erosion by the

---

[2]Webi, in the Somali language, means "river".

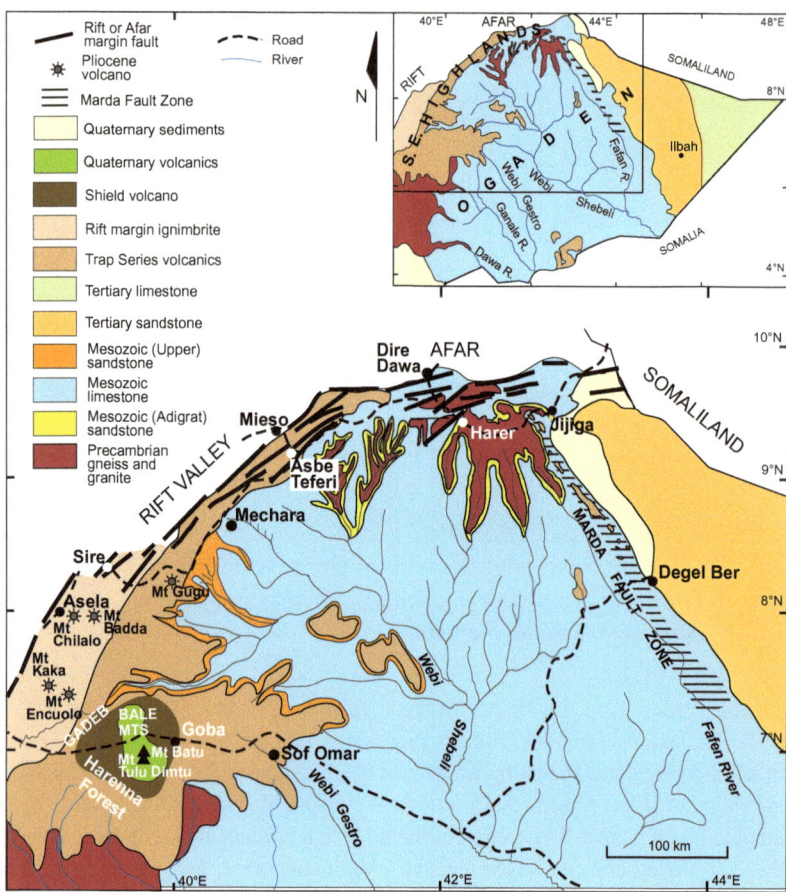

Fig. 15.3 Geological map of the region shown in Fig. 15.2. The *inset* shows the whole area covered by the Southeastern Highlands and the Ogaden; the *rectangle* outlines the region shown in the main map. Modified from Geological map of Ethiopia 1:2,000,000 2nd Ed. (1996), Geological Survey of Ethiopia

Webi Shebeli's tributaries is slowly cutting away the southern edge of these mountain ranges.

The highlands' cover of volcanic rocks thins out to the east along the Ahmar Mountains and eventually disappears altogether. Near the top of the descent to Dire Dawa, which lies at the foot of the mountains and on the edge of Afar, Precambrian gneisses and migmatites are exposed (PHOTO 15.1). At the other end of the

PHOTO 15.1 Precambrian migmatite ("mixed rock") in a road cutting near the top of the descent to Dire Dawa Field notebook for scale (2014)

highland ridge, southwest of the Chercher Mountains, is a broad and rather complex region of volcanic rocks (Fig. 15.3). Four imposing volcanoes, Kaka, Badda, Encuolo and Chilalo, of Pliocene to early Pleistocene age (about 2 million years old) and formed of trachyte, border the Rift Valley and are thought to be associated with the later stages of rift margin faulting. Eastward of these volcanoes is an expanse of Trap Series volcanics surmounted by the magnificent Bale Mountains, remnant of a large shield volcano.

15.1 The Bale Mountains

In contrast to the Western Highlands, which are characterised by several big shield volcanoes of Miocene age, the Southeastern Highlands have just one: the Bale Mountains. Like the Semien, the Bale Mountains are referred to in the plural because of the numerous peaks which are present. They are best known and most visited for their populations of endemic animals such as the Ethiopian Wolf and

PHOTO 15.2 Part of a moraine ridge in the Bale Mountains. The piled-up rocks were carried and dumped by a glacier, possibly around 20,000 years ago although the moraine itself has not been dated (2010)

PHOTO 15.3 Garba Guracha, a cirque lake in the summit region of the Bale Mountains. The cirque, or hollow, in which the lake is contained, was carved by ice at the head of a glacier (2010). Photo courtesy Emile Farhi

Mountain Nyala, their birds and alpine plants, and the opportunities they offer for hiking and horse-riding. Indeed, one of the first things that will strike the visitor to the Bale Mountains is that there are horses everywhere, many of them colourfully decorated with ribbons and strings of red rosettes. The Oromo people of the region are great horsemen, and for the visitor the back of a horse is an ideal way to explore the beautiful alpine scenery and striking geological features of these mountains.

Unlike the Semien, which are built mainly of basalt, the Bale Mountains are formed largely of trachyte lavas. These appear to have erupted from several vents, so the edifice is quite a complex one. The bulk of the lavas probably erupted between about 30 and 7 million years ago, although I have not been able to find any measured ages. There have certainly, however, been some more recent eruptions. For example, a rock sample from Mt Batu, the second highest peak, has been dated at only two and a half million years old.

Although comparable in age, expanse, elevation and origin to the Semien, the scenery of the Bale Mountains is markedly different, though no less impressive. Their highest peak Mt Tulu Dimtu is, at 4400 m, only about 150 m lower than that of Ras Dashen, yet they lack the dramatic gorges and sheer escarpments of the Semien. This is because they have been far more modified by glaciation. Around 22,000 years ago, during the last great ice age when much of Europe and North America was ice-bound, ice covered an area of about 180 km^2 in the Bale Mountains. A 30 km^2 ice cap covered and surrounded the peak of Mt Tulu Dimtu, and individual glaciers reached down to 3200 m. The glaciers have left evidence in many forms: for example the U-shaped valleys which they scoured out, boulders scratched by their bedload of rocks, moraine ridges of the rock debris they carried and deposited, and depressions known as cirques where ice accumulated and excavated circular basins (PHOTOS 15.2, 15.3 and 15.4). The glaciers have rounded out the topography, resulting in isolated peaks separated by broad valleys, and revealing the interiors of volcanic vents as plugs (PHOTO 15.5). They have essentially worn away much of the original volcano and exposed its roots.

At first it may seem puzzling that the Bale Mountains were glaciated to a much greater extent than the Semien which, besides reaching a slightly greater elevation, are also at a higher latitude. The answer lies in a difference in climatic conditions. Water as well as cold is required to produce glaciers, and rainfall then was higher in the region of the Bale Mountains than in the Semien. Not only the Bale Mountains but also the nearby peak of Badda, and possibly those of Kaka, Chilalo and Encuolo, were glaciated at that time.

The Bale Mountains are, like the Semien, partially bounded by an escarpment. This curves around the southwestern and southeastern sides of the massif. As with the Semien, it is the result of headward erosion, in this case by tributaries of the

PHOTO 15.4 The scratches, or striations, on this boulder were caused by a glacier passing over it, containing rock fragments which acted as an abrasive (2010)

PHOTO 15.5 A volcanic plug in the Bale Mountains summit region, framed by giant lobelias (2010)

Ganale River. At the foot of the southwestern escarpment is the beautiful Harenna Forest. This is Ethiopia's second-largest natural forest—a dense, green, misty jungle of huge trees (fern pine, fig, juniper), moss-draped branches, leaping colobus monkeys and an impenetrable undergrowth wrapped in a tangle of creepers among which wild coffee grows.

To the west of the mountains a smaller and gentler escarpment slopes down to a plain known as Gadeb, in the uppermost headwaters of the Webi Shebeli. Part of this plain is the floor of a lake that dried up more than 2 million years ago. Prehistoric people camped by the rivers which subsequently flowed across the dry lake bed, leaving behind not only their stone tools but also evidence that they may have used fire. Magnetic minerals within a number of burned rock fragments have been found to be aligned with what would have been the earth's magnetic field direction at that time, about one and half million years ago, something which could only happen if they had been heated to a very high temperature. This would be one of the earliest instances of the use of fire by humans.

East of the Bale Mountains, southeast of the Chercher Mountains and south of the Ahmar Mountains the Trap Series lavas thin out, giving way to the great limestone plain of the Ogaden which we will come to later in this chapter. The headwaters of the Webi Gestro and Webi Shebeli have cut their way back into these highlands and carved a low escarpment along their southeastern edge, often dissected by deep gorges (PHOTO 15.6), and exposing the underlying limestone. In the limestone are numerous caves, many of which are unexplored and doubtless many still to be discovered.

15.2 The Caves of Sof Omar and Mechara

The best-known of these caves, just east of the Bale Mountains, is the magical underworld of Sof Omar. Here an intricate system of underground passages and caverns, over 15 km long, constitutes the longest cave system in Ethiopia and possibly in the whole of Africa. As well as being of outstanding beauty, it is a shrine of great cultural and religious significance to the surrounding Islamic and Oromo people.

It began forming many millennia ago, as water slowly seeped through joints and cracks in the limestone, dissolving it and in places causing it to disintegrate and collapse. A network of passages formed, some just narrow cracks, others broad passageways widening to great domed chambers. At some stage the Webi Gestro, rising east of Goba in the Bale Mountains (see Fig. 15.3), abandoned its above-ground course to flow underground through these passages, further

PHOTO 15.6 The Webi Shebeli cuts a gorge through limestone on the southern side of the Southeastern Highlands, near to Sheik Hussein (2014). Photo courtesy Peter Purcell

widening them by erosion. The old, abandoned river course can still be seen as a dry valley. Although the presence of water within the cave system renders exploration more challenging (a certain amount of swimming is required), it further enhances its beauty. The spectacular Chamber of Columns, for example, resembles a cathedral whose pillars, carved to intricate shapes by the rise and fall of the water, are reflected in a still pool (PHOTO 15.7). These pillars, and the other formations in Sof Omar, are entirely the work of water flowing through the system. Decorations in the form of stalactites or stalagmites are absent.

Most of the caves in this region occur, like Sof Omar, in the headwaters of rivers flowing from the Southeastern Highland ridge, but rarely do streams pass through them as in the case of Sof Omar. Unlike Sof Omar, some are abundantly decorated with stalactites and stalagmites, and at a group of caves near Mechara these are being studied in order to determine how their growth and chemical properties are related to climate. One may wonder why these caves contain such decorations whereas Sof Omar does not. It is perhaps to do with the fact that much of the Sof Omar system still retains a basalt capping which inhibits the constant, slow percolation of water whose dripping is responsible for the formation of

PHOTO 15.7   The Chamber of Columns in the Sof Omar caves. Photo courtesy Eugenio Lizardi

PHOTO 15.8   Rock art from Dessa Cave (Goda Dessa), about 40 km west of Dire Dawa. This painting is particularly interesting as it depicts both wild and domesticated animals. Although the age of the painting is not known, it was probably done sometime during the last 10,000 years. Photo courtesy Zelalem Assefa

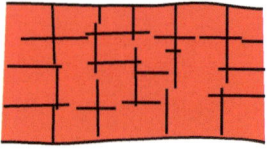 (i) The rock cracks as material above it is eroded away, relieving the pressure on it. This results in a roughly rectangular pattern of joints.

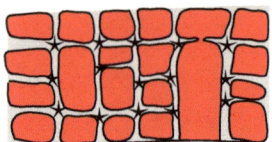 (II) Water seeps along the joints, gradually breaking up the rock material. Corners are more readily attacked than faces and edges, resulting in rounded boulders.

 (iii) Eventaully the boulders become so worn that they can no longer support each other, and collapse into a heap. A few may still remain standing, balanced upon each other.

Fig. 15.4 How rocks such as granite and granite gneiss weather to formations like those seen in the Valley of Marvels

decorations. At Mechara the limestone is directly covered by soil and has no basalt capping so that rainwater can more easily percolate into the caves.

Many of the caves and rock shelters of the Southeastern Highlands were occupied by prehistoric peoples. They left behind their stone tools, and evocative paintings depicting human figures together with domesticated animals such as sheep and cattle, as well as the wild animals with which they would have been familiar (PHOTO 15.8). Judging from the paintings, these last were far more varied and abundant when those artists lived than they are at present. The paintings have not been dated, but would have been done during the past 10,000 years since prior to that people did not have domesticated animals.

15.3   The Valley of Marvels

East of Harer the limestone thins out, and the underlying Precambrian basement is exposed. The basement here is formed of granite and granite gneiss (a metamorphic rock similar in composition to granite, but with a banded texture), types of

PHOTO 15.9 Precariously balanced boulder of granite-gneiss in the Valley of Marvels (Dakhata Valley), between Harer and Jijiga (2014)

rock which tend to weather into rounded boulders by a process illustrated in Fig. 15.4. These boulders can remain piled on top of each other for hundreds, or perhaps thousands, of years, often after having become quite disconnected. This results in some extraordinary and spectacular rock formations as seen in the Dakhata Valley, also known as the Valley of Marvels, about halfway between Harer and Jijiga (PHOTO 15.9).

15.4 The Marda Fault Zone and the Marda Pass

Just west of the town of Jijiga, and east of the Valley of Marvels, is a line of low hills (PHOTO 15.10) known as the Marda Range. The hills are formed of limestone, capped by basalt, and are part of a linear feature which extends in a northwest to southeast direction across most of the region shown on Fig. 15.2. Geologically this feature is known as the Marda Fault Zone. It is a broad band of

PHOTO 15.10 Looking westward toward the Marda Range, on the skyline, with the town of Jijiga in the foreground (2008). Photo by Peter Purcell. Fig. 19.3(a), Billi (2015) by permission of Springer

faults, in some ways similar to the Yerer-Tulu Welel Volcanotectonic Lineament which we met in Chap. 12, although volcanic features are less apparent. On high-resolution satellite images, and using geophysical methods such as gravity measurements, the Marda Fault Zone can be traced much further to the southeast, traversing the Ogaden plain and extending almost as far as the coast of Somalia. To the northwest it aligns with the volcanic ranges in northern Afar, and then with the coast of the Red Sea. This interesting feature clearly represents a major geological structure, following a line of weakness inherited from the ancient continental collision of Precambrian times.

The road between Harer and Jijiga passes through a gap in the range, known as the Marda Pass. This pass has some interesting historical connections. It was, for example, the site of major fighting during the defeat of the Italian army by British and Ethiopian forces in 1941, and a strategic strongpoint during the war between Ethiopia and Somalia in 1977–1978.

15.5   The Ogaden Plains

The Ogaden, named after one of the numerous clans who inhabit it, is not a formally defined region politically, ethnically or geographically. Since it has no official boundaries it is hard to say just where it starts and ends, and some of the regions described in the previous sections: the limestone caves and certainly the Marda Range, may be considered as belonging to it. In this section we will look at the vast plain which constitutes its greater part, extending from beyond the Southeastern Highlands to Somalia. Hot, dry, and sparsely populated, it is largely a landscape of flat scrub, punctuated by occasional small, bare volcanic hills. Unfortunately it has long been a bone of contention between Somalia and Ethiopia, and amongst the various Somali clans who occupy it, with the result that security issues have restricted access to it for many years. This is a pity since, in addition to its starkly compelling scenery and some very interesting geological features, it potentially contains reserves of crude oil and natural gas which would be a great boost to Ethiopia's economy.

Back at the end of Palaeozoic times, as we saw in Chap. 6, a broad rifted basin formed across the future Ogaden. Mesozoic rivers flowed across it, depositing their loads of gravel and sand as they had in the Blue Nile basin to the northwest. Although far from the Blue Nile, and even further from Adigrat itself, this sandstone is assigned to the Adigrat Sandstone Formation as it was deposited around the same time, under similar conditions, and has the same distinctive characteristics of composition and structure as the sandstone at those localities. Following the deposition of the sand, the sea advanced from the southeast to cover the whole of the Ogaden and eventually much of Ethiopia. It took a long time to retreat from the Ogaden region, which was slowly subsiding as the Indian Ocean opened up. In fact, as noted in Chap. 6, it made several advances and retreats over the region during the Cretaceous period, and even the early Tertiary, before it finally retreated completely. This region was therefore covered by the sea for much longer than anywhere else in Ethiopia and great thicknesses of limestone, shale and gypsum, possibly exceeding 3 km, had time to accumulate.[3]

The main limestone formation corresponds to the Antalo Limestone which we have already seen in the Blue Nile Gorge and in Tigray. In the Ogaden, however, because it is thicker and was deposited over a longer period of time, it has been given a different name and is known as the Hamanlei Limestone. Overlying this limestone, and interbedded with it, are layers of shale—compressed and

---

[3]The whole region of limestone and other sedimentary rocks which underlie the Ogaden is known as the Ogaden Basin, not to be confused with the ancient rifted basin of Palaeozoic and early Mesozoic times.

consolidated mud mixed with decayed organic material. Under the pressure and heat of burial, this material has generated oil and gas which seeped into porous layers in the limestone and the underlying sandstone, and were trapped by impervious layers above. A considerable amount of exploration has been undertaken here by various oil companies, revealing a promising resource, but until the region becomes politically stable this must remain hidden beneath the dusty plain of the Ogaden.

You can see on Fig. 15.3 that the Marda Fault Zone marks a divide between quite different geological formations in the Ogaden plain. To the southwest are Mesozoic limestones and to the northeast Tertiary sandstones and Quaternary sediments. The scenery on either side of the fault zone reflects this difference, as the limestones form a rough, bumpy surface with an overall greyish appearance while the sandstones weather to a sweeping red sandy plain. The fault zone may have acted as a kind of hinge, with the land to the east of it sagging to form a coastal basin as the sea withdrew. Sand was deposited both along the shoreline of the retreating sea, and by rivers flowing from the land to the north and west.

PHOTO 15.11   This hill near Ilbah village, in the sandy plains of the eastern Ogaden, is the exposed top of a basalt dike. The hill is oriented north-south, and the photo is taken from the air, looking toward the southwest (2008). Photo courtesy Peter Purcell

PHOTO 15.12 Meanders of the ancestral Webi Shebeli, now depicted by a flow of basalt which has occupied the ancient valley. If you look carefully, the flow can be seen extending as far as the skyline in a huge meander (2008). Photo courtesy Peter Purcell

Further to the east the sandstone is covered by limestone resulting from at least one further incursion of the sea.

Four major rivers flow across the Ogaden plain: the Ganale, Gestro, Shebeli and Fafan. The longest of these is the Webi Shebeli whose name, meaning "the river of leopards", captures the imagination even though it is doubtful that there are still many leopards in the region. Early Arab cartographers, recognising its great length and importance, termed it the Second Nile. It is a vital waterway in this arid region, and for many decades has enabled crops of maize, sesame, fruit and vegetables to be cultivated in the vicinity of its banks. At the Somali border the Ganale and Gestro join the Dawa River to form the Juba which, like the Shebeli, eventually reaches the Somali coast.

In Ethiopia one can never get far from volcanics, and the Ogaden is no exception. Here there are two particularly interesting volcanic formations: basalt hills and basalt rivers! The small basalt hills which are dotted mainly over the eastern part of the Ogaden (PHOTO 15.11) represent dykes—narrow intrusions of basalt magma which almost but not quite reached the surface. Their tops have since been exposed by erosion of the material around them. Individual ones are too small to show on Fig. 15.3, but there are so many of them, both visible as hills or detected by magnetic surveys, that they have been described as a swarm. They have been dated at about 25 million years old and are part of the volcanic activity which heralded the break-up of Africa and Arabia.

The basalt "rivers" are not truly rivers—but are basalt lava flows which have occupied the old valleys of real rivers. Because lava is fluid it will follow a channel if one is available, sometimes displacing the river that is already in it and forcing it to follow a new course. PHOTO 15.12 shows an example of a basalt flow which has followed an old channel of the Webi Shebeli. Because the basalt is more resistant to weathering than the surrounding sediments it is left standing as they are eroded away. This is an example of what is termed "inverted topography" where something that was originally a valley is preserved as a ridge—an upside-down valley. This particular basalt-filled valley can be traced for over 200 km—and the modern Webi Shebeli follows a course nearby.

More detailed observations of the Ogaden's intriguing geological features must, like the oil exploration projects, await the cease of hostilities and return of security to the region.

Introduction to the Rift Valley

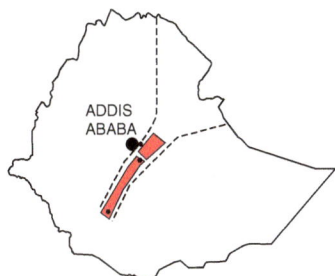

We have seen in previous chapters that Ethiopia is cut diagonally by the Ethiopian Rift Valley, a low-lying strip of land that separates the Western and Southeastern Highlands, and it is now time to take a closer look at this remarkable feature. Together with its extension northwards into Afar, it is probably the most exciting and significant geological feature in Ethiopia.

Before we start, it is useful to clarify just what is meant by the terms "rift" and "rift valley". In everyday life a "rift" implies some kind of break with a gap in between, such as a rift in one's relationship with a friend. In geology it means much the same thing. It is a break in the earth's surface, where a strip of land has subsided between pairs of cracks, or faults. The term "rift valley" is generally used synonymously with "rift" but emphasises the valley-like nature of the feature. In an ideal rift valley the land has subsided between two opposing sets of faults. In reality, however, some rift valleys are bounded by a fault, or series of faults, on one side and by a flexure or just a slope on the other, and there are many varieties in between. Figure 16.1 illustrates just a few of the profiles a rift valley may have.

The Ethiopian Rift Valley and Afar are just one section of a much bigger system of rift valleys, faults and sea-floor spreading axes, in total over 6000 km long, which extends from northern Syria to the coast of the Indian Ocean in Mozambique. It is basically a long and rather messy tear in the earth's crust, representing the latest stage in the break-up of the Gondwana supercontinent.

© Springer International Publishing Switzerland 2016
F.M. Williams, *Understanding Ethiopia*, GeoGuide,
DOI 10.1007/978-3-319-02180-5_16

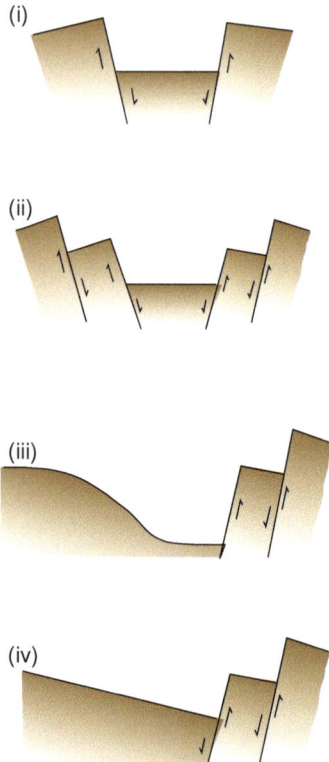

Fig. 16.1 Some examples of rift valley profiles. Rifts having profiles (**iii**) and (**iv**) are sometimes referred to as "half graben", as they are bounded by a fault on one side only. (**i**) Simple rift. (**ii**) Rift with multiple faults. (**iii**) Rift with fault(s) on one side and a flexure on the other. (**iv**) Rift with fault(s) on one side and a slope on the other

Figure 16.2 shows the whole of this rift system and illustrates its different sections. Those sections that occur within the African continent are known as the African Rift System. At its northern end, in Afar, the African Rift System connects with the Red Sea and the Gulf of Aden, along whose central axes new ocean crust is being created as the Arabian Peninsula swings away from Africa. The Red Sea itself connects, via the Gulf of Aqaba, with the Dead Sea Rift in the Middle East. Here the Arabian Plate is sliding northwards against the African Plate and colliding with the Eurasian Plate, to form the Zagros mountain range which curves through Iran.

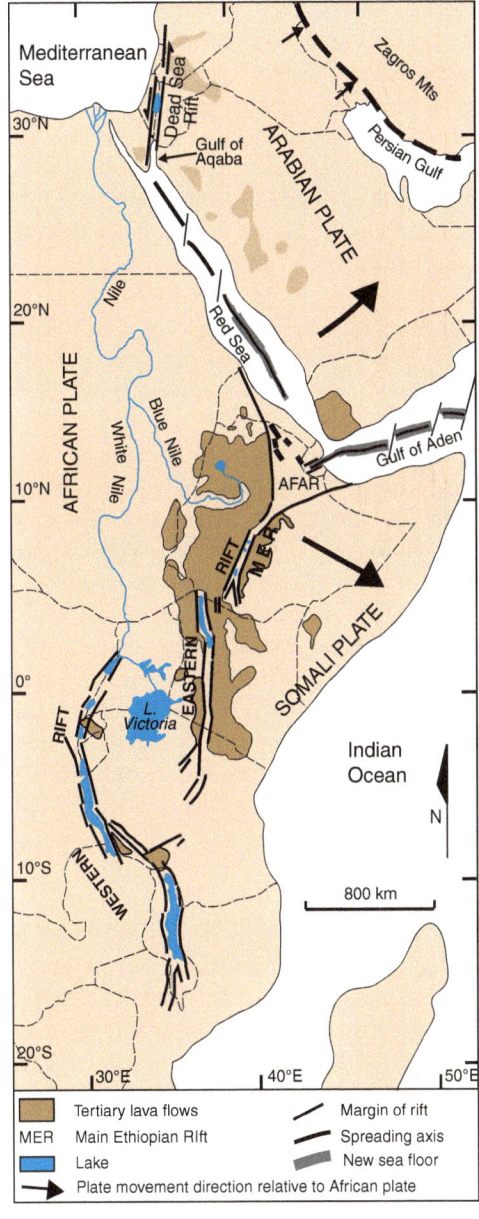

Fig. 16.2 The rift system

The Gulf of Aden and the Red Sea are widening at a rate of 1.5–2 cm per year. New sea floor is being produced along their spreading axes and they are on their way to becoming true oceans. The African Rift System is quite different. With the exception of Afar, it is underlain entirely by continental crust, no new sea floor is being produced and it is widening at no more than half a centimetre a year. Whether it will continue to widen and eventually become an ocean is a matter for conjecture, which we will consider further in Chap. 26.

The African Rift System consists of a series of disconnected rift valleys which include all the types shown in Fig. 16.1, as well as many complex variations on them. As a whole it marks an untidy break in the African Plate, as present-day Somalia, the eastern parts of Ethiopia, Kenya and Tanzania, and the northern part of Mozambique pull away from the rest of the African continent. The reason the break is so untidy is that it is trying to follow weak zones where the crust had been sheared, folded and fractured during the continental collision of the East African Orogeny, and avoid the stronger blocks that had remained intact.

South of Afar the African Rift System is made up of two segments—the Eastern Rift and the Western Rift. The Eastern Rift, of which the Ethiopian Rift Valley is part, runs roughly north-south from Ethiopia to central Tanzania. The Western Rift curves around the western side of Lake Victoria, passing through Uganda, Rwanda and Burundi, then continues southwards around the edge of Tanzania and through Malawi into Mozambique. Lake Victoria, between the two rifts, is situated in a depression within a very solid block of crust, which is why they pass around it. Note that the tear in the earth's crust is actually much messier than Fig. 16.2 indicates. Almost every fault-line shown consists in reality of a complex of smaller faults, some parallel and some even criss-crossing each other. The figure just gives a general and very simplified picture.

Within Ethiopia the rift system has two distinct sections: Afar and the Main Ethiopian Rift. The complex and fascinating region of Afar will be visited in Chaps. 21–23; in Chaps. 17 and 19 we will look at the Main Ethiopian Rift.

The Main Ethiopian Rift itself may conveniently be considered in three sections, as shown in Fig. 16.3. The northern section extends roughly from Awash to the town of Nazret (Adama), the central section from Nazret to Lake Awasa, and the southern from Lake Awasa to Lake Chamo. The divisions are not of course sharply defined, but neither are they entirely arbitrary. The northern section is a transition region between the Main Ethiopian Rift and Afar. The boundary between it and the central section is marked by a change in the orientation of the rift, from NE-SW to NNE-SSW, and a noticeable change in the rock types forming its floor. In its central section the rift is very clearly defined—almost as ideal as a rift can be! Its boundary with the southern section is marked by a slight change in orientation

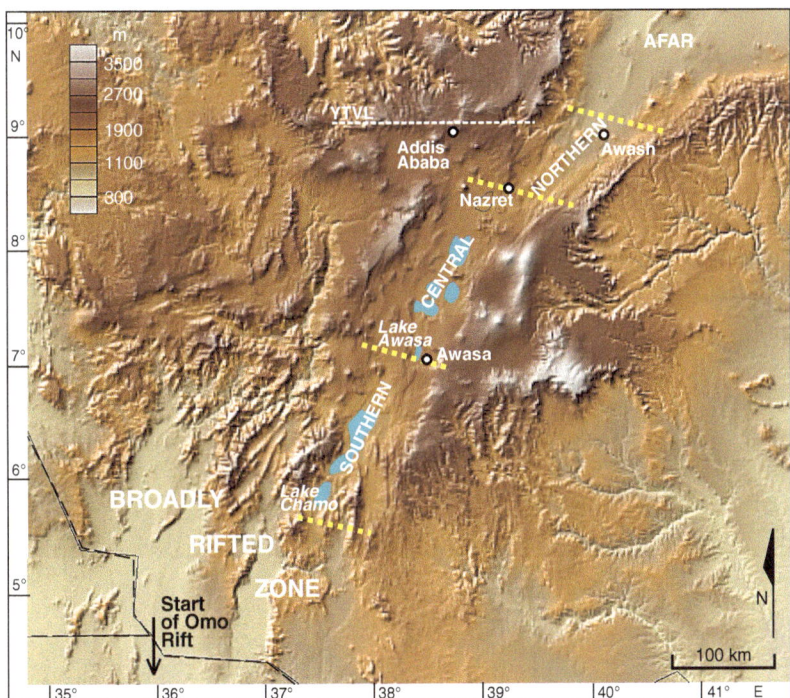

Fig. 16.3 Digital elevation model of the Main Ethiopian Rift, indicating its different sections. The *yellow dotted lines* demark the sections. The *white dashed line* labelled YTVL is the Yerer-Tulu Welel Volcanotectonic Lineament which was discussed in Chap. 12. DEM from GeoMapApp

to a more southerly direction, and its western margin becomes less clearly defined. South of Lake Chamo the rift ceases to be a distinctive feature at all. It breaks up into a number of faults and ill-defined rifts, and for this reason the region is sometimes referred to as the Broadly Rifted Zone. The Eastern Rift recommences as a distinct feature with the Omo rift in northern Kenya, from where it continues southwards through Kenya and into Tanzania.

In order to understand the features that we will see in the Main Ethiopian Rift, we need to take a look at how it formed. This is summarised in Fig. 16.4. Each section of it started as a sag in the earth's crust which, being brittle, eventually fractured when it could bend no further. Strips of crust slipped down between the fractures to form rifts. This was accompanied by different styles of volcanic

(i) Forces pulling the African Plate apart stretch the crust, which thins a little and sags, but because it is brittle it cannot stretch very much before it begins to crack. (~15-10 My ago)

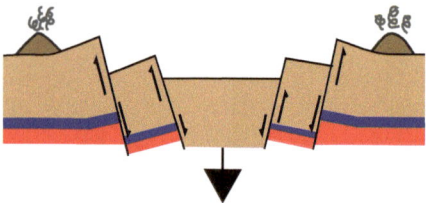

(ii) The edges of the sag break completely and become the rift margin faults, The central strip slips down between the faults. Large volcanoes form close to the rift shoulders. (~10 - 2 My ago)

(iii) The volcanoes on the rift shoulders cease activity, but explosive eruptions accompany the rift faulting as hot molten material bursts through, buoyed up by gases into a fiery froth. The centre part of the rift continues to drop between the faulted margins, into the space vacated by the molten material. The volcanic froth rolls across the rift floor where it cools and solidifies as ignimbrite. (~2-1 My ago)

(iv) Volcanic activity along the rIft margin faults, and movement of the faults, subside. Instead the rift floor itself begins to crack, and a band of faults and fissures form along it. Large silicic central volcanoes erupt along the rift floor. The volcanic and tectonic activity has moved inwards from the rIft margins to within the rift itself. (~1 My to present)

Fig. 16.4 Stages in the formation of the Main Ethiopian Rift. The time ranges of events, given in My (millions of years before the present), are only very approximate, as events overlapped and occurred at slightly different times in different parts of the rift

activity, which we will look at more closely in the next chapter. Finally, the volcanic activity and the fracturing process moved inward from the margins of the rift to the rift floor itself, and this is the stage it is at today.

Clearly this whole process involved dramatic events aplenty, but to put them into perspective it is important to remember that the drama was enacted over a period of more than 10 million years. Had there been rift valley dwellers at the time the initial sagging was taking place, they would have been unaware that anything was happening. The formation of the rift margin faults would have been accompanied by earthquakes and occasional but violent volcanic activity. During this period, which took place over 4–9 million years, our hypothetical inhabitants would probably have experienced occasional earth tremors, often quite strong, while volcanic eruptions would be rare but devastating events. By about a quarter of a million years ago, when the activity had moved to within the rift, people may have been living there. For them it may have been rather like living somewhere such as New Zealand or Iceland today, where a person might be aware of several earthquakes, and perhaps a volcanic eruption, during their lifetime. Today the activity has quietened down a lot, but occasional earth tremors, the sudden appearance of cracks and fissures, and ubiquitous hotsprings and fumaroles (steam vents) show that it has by no means ceased.

The Red Sea and the Gulf of Aden would have been born in much the same way as the Main Ethiopian Rift, but have evolved further to develop fully-fledged spreading axes along which ocean floor is being produced.

We will begin our visit to the Ethiopian Rift Valley in the middle, with the central and southern sections of the Main Ethiopian Rift, then move southwards to the Broadly Rifted Zone and finally, in Chap. 19, to the northern section which will link us to Afar and the climax of Ethiopia's geological story.

The Rift Valley Southwards: Volcanoes, and Lakes Ancient and Modern

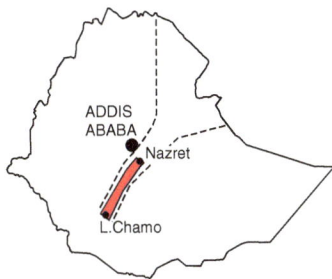

The Ethiopian Rift Valley is best known for its beautiful lakes with their abundant bird and other wildlife and, in recent years, pleasant lakeside lodges ranging from the comfortable to the luxurious and offering a range of recreational activities. Few visitors, however, are aware that they are standing upon a unique feature of the earth's surface and that beneath their feet a continent is in the early stages of breaking apart. To the geologist the Rift Valley, with its many manifestations of this process, provides an irresistible outdoor laboratory for furthering understanding of how our earth works.

In this chapter we will explore the central and southern sectors of the Main Ethiopian Rift,[1] as designated in the previous chapter. Together they extend roughly from Nazret (Adama) to Lake Chamo, and for the most part are bounded on both sides by escarpments which rise to about 1500 m above the rift floor. From

---

[1]The term "Main Ethiopian Rift" refers to that section of the Rift System between just north of Awash, and Lake Chamo. The terms "Ethiopian Rift Valley", "Ethiopian Rift", "Rift Valley" and simply "Rift" are also used, but sometimes these implicitly include Afar. This is one of many instances where terminology is used loosely and can lead to confusion. In this book, use of any of the above terms applies to the Main Ethiopian Rift only. Afar will always be referred to by name.

© Springer International Publishing Switzerland 2016
F.M. Williams, *Understanding Ethiopia*, GeoGuide,
DOI 10.1007/978-3-319-02180-5_17

PHOTO 17.1 Looking westward across the flat floor of the Rift Valley to the Gurage Mountains, almost merging with the clouds in the background and forming the main rift escarpment here. In front of the escarpment is a line of low volcanic hills, near Butajira (2014)

within, it is almost always possible to see one or other of the bordering escarpments (PHOTO 17.1), though it requires an unusually clear and dust-free day to see both simultaneously. Since the rift here is some 80 km wide, far wider than it is deep, a traveller may not have the sensation of being in a trough-like valley but they will certainly feel that they are in a different world. The flat rift floor, punctuated by volcanic cones and fresh fault scarps, and the warm, almost soporific, climate could hardly be in greater contrast to the rugged topography and cool freshness of the adjacent highlands.

The main geological features of the central and southern sections of the Main Ethiopian Rift are shown in Fig. 17.1. You will notice immediately that most of the

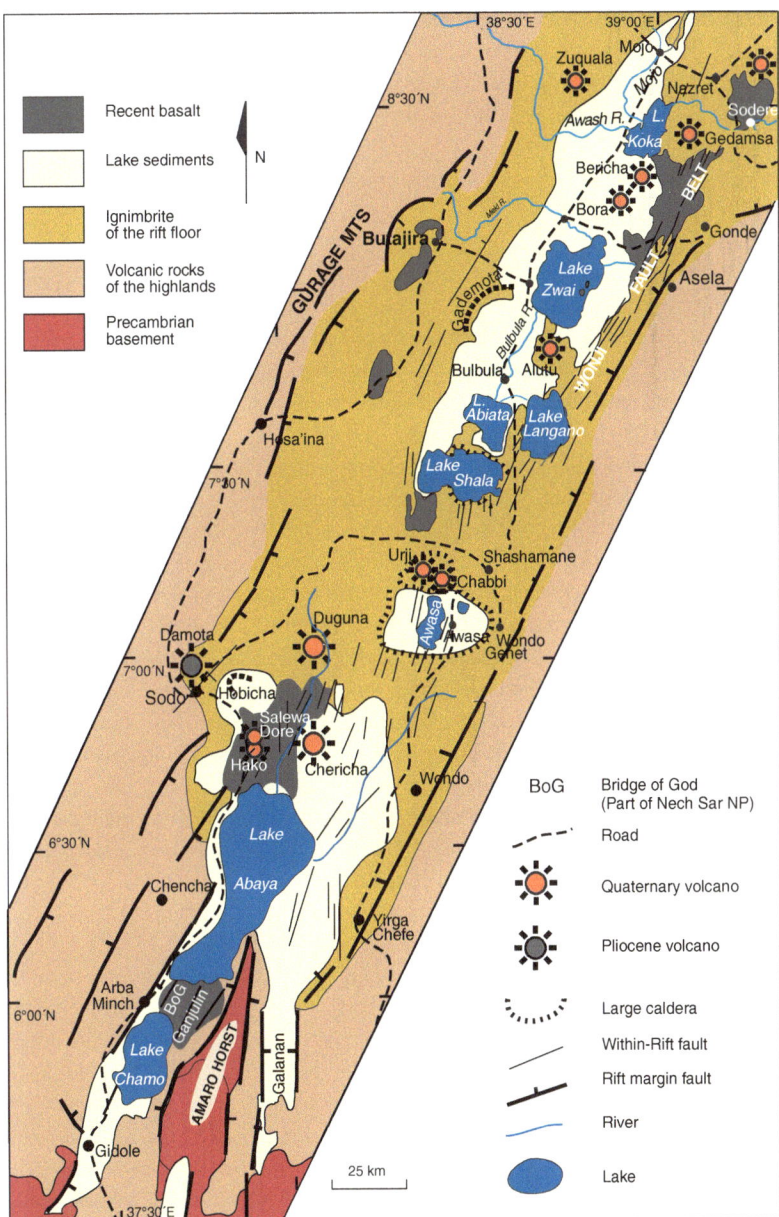

Fig. 17.1 Geological map of the Main Ethiopian Rift between Nazret and Lake Chamo. Modified from Geological map of Ethiopia 2nd Ed. (1996) and Geological maps of Hosana, Dila and Hagere Mariam 1:250,000, Geological Survey of Ethiopia

rift floor is covered by two materials: ignimbrite, shown in mustard yellow, and lake sediments, shown in pale yellow. Ignimbrite has been mentioned in previous chapters, but it is such an important and abundant rock in this part of the rift that it is time to take a closer look at it.

17.1 Ignimbrite

Ignimbrite is a rock formed as a result of an extremely violent and explosive volcanic eruption. Eruptions of this kind are due to the sudden release of pressure on magma (molten rock) which contains a lot of gas. It is rather like popping the cork of a champagne bottle—the liquid gushes out in a bubbling mass as the pressure on the gas it contains is released. In the geological situation the liquid is white-hot molten rock, and the gas is a mixture of water vapour, carbon dioxide, sulphur dioxide, hydrogen sulphide and other noxious compounds. The cloud of gas and liquid droplets bursts from the source of the eruption carrying a load of fine ash, crystals, and rock fragments which it has picked up on the way, and rolls away over the surrounding land. This dense, white-hot cloud is known as a nuée ardente ("burning cloud") and is the most dangerous and destructive type of volcanic eruption, since the cloud can reach temperatures of 1000 °C and travel at speeds of up to 700 km per hour, destroying everything in its path. As it eventually cools and halts, the whole mass becomes compressed under its own weight. The resulting rock is ignimbrite, from the Latin *igni*: "fire" and *imbri*: "rain". There are various kinds of ignimbrite. Sometimes sufficient heat is present during the settling process to fuse the solid fragments together, in which case it is called a welded tuff. Ignimbrite containing a large quantity of pumice fragments is referred to as pumiceous ignimbrite, or sometimes as pumice tuff. Whatever the case, ignimbrite is quite an easy rock to recognise on account of the fragments it contains. PHOTO 17.2 shows a pumiceous ignimbrite, and if you look ahead to Chap. 19, PHOTO 19.3 shows a very beautiful welded tuff.

Nuée ardente eruptions can occur through a fissure, or via a volcanic vent. The rift ignimbrites have been produced in both ways. A great deal of ignimbrite activity was associated with the faulting which formed the rift margins. Molten material burst through the cracks, pouring over the newly forming margins onto the rift floor, and the removal of this material from below created space into which the rift floor subsided. Much ignimbrite has also been produced by the volcanoes along the rift floor. Often the force of the explosion was sufficient to blow the entire top off the volcano, which then collapsed to form a very large crater called a caldera (Fig. 17.2).

PHOTO 17.2 Pumiceous ignimbrite near Lake Langano. Note the many small rock fragments. There are also many stretched-out streaks of pumice, which appear slightly *grey*. They are not easy to see in the photo, but there is one just left of centre. 20c piece for scale (2008)

Ignimbrite underlies almost the entire floor of the central, and much of that of the southern, Main Ethiopian Rift, to depths up to at least 400 m. Even the pale yellow areas shown on Fig. 17.1, the lake sediments, have ignimbrite beneath them. Basalt, on the other hand, which is by far the dominant rock type in the highland regions, occurs as only a few scattered flows in this part of the rift. Most of these flows look very fresh, and they are the youngest rocks in the rift. The flows have erupted through fissures, and are often associated with the small black cinder cones that dot the rift floor. These cones are themselves built of basalt but in the form of cinders—blobs of lava filled with gas bubbles.

17.2 The Rift Volcanoes

Running along the rift floor, slightly to the east of its centre, is a line of prominent volcanoes (PHOTO 17.3). These are shown in orange on Fig. 17.1. Actually Lake Shala is also a volcano, but its top has completely collapsed to form a caldera

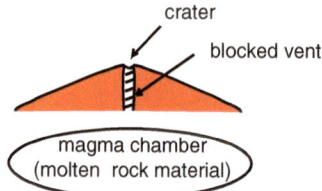

1. A concentration of hot, molten rock material (magma) is lying
beneath the volcano, waiting to burst through. The space where
the molten material is confined is known as a magma chamber.

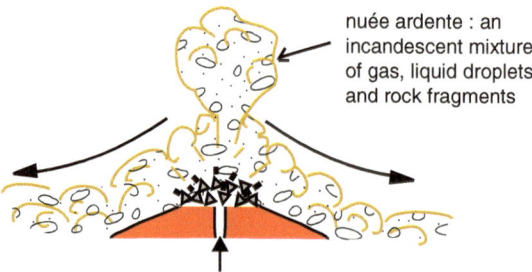

2. The magma finally bursts through the blocked vent in a great explosive
eruption, producing a fiery cloud (nuée ardente) of gas , liquid
droplets and rock fragments, and blowing off the top of the volcano.

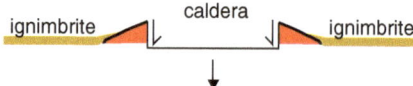

3. What is left of the volcano's summit collapses
into the emptied magma chamber. The nuée
ardente solidifies as ignimbrite.

Fig. 17.2 How a caldera is formed

which is filled by a deep lake. All these volcanoes are dormant, still showing signs
of activity in the form of hotsprings and fumaroles. Their lavas are mostly less than
a million years old and, judging by their fresh appearance, their most recent
eruptions probably occurred only thousands, or even hundreds, of years ago.

PHOTO 17.3   Alutu volcano, between Lakes Zwai and Langano. The top of the mountain appears flat because it has collapsed to form a caldera. The caldera itself is now mostly obliterated by flows of pitchstone (2007)

Resourceful local people have put the hotsprings and fumaroles to numerous uses, from washing and cooking to saunas and therapeutic baths (PHOTO 17.4). They are also being investigated as a potential source of geothermal energy, and the geothermal power station near to Alutu volcano should be fully in operation by the time you read this book.

The rift volcanoes are formed of silica-rich rocks such as rhyolite and trachyte. The beautiful rock obsidian, a black, shiny volcanic glass, and its near relative, pitchstone, are also abundant (PHOTO 17.5). Obsidian is formed from silica-rich lava that cooled too rapidly for any crystals to form, whereas pitchstone contains some crystals and is less shiny. Pumice, a light, frothy rock which consists essentially of holes connected by glassy threads, is also common. The volcano Chabbi, near to Shashamane, is constructed entirely of pumice and obsidian; nearby Urji, and Bericha north of Lake Zwai, are made almost wholly of pumice.

PHOTO 17.4   An improvised sauna makes use of a fumarole on Chabbi volcano (2008)

PHOTO 17.5   Part of an obsidian flow on Chabbi volcano. Geological hammer for scale (2008)

Most of these rift volcanoes have calderas and have erupted large quantities of ignimbrite. Shala is almost entirely caldera, with virtually none of its original volcanic edifice remaining. Some older caldera volcanoes, for example Gademota, have been partially or completely buried by later eruptions of ignimbrite.

17.3 The Rift Lakes

Although the lakes of the Main Ethiopian Rift: (Zwai, Langano, Abiata, Shala, Awasa, Abaya and Chamo) are tiny in comparison to the great lakes of the East African countries further south (Albert, Edward, Tanganyika, Malawi and Victoria), they are no less beautiful and are fascinating in their diversity. They also have much to offer in the way of geological interest.

Lakes Zwai, Langano, Abiata and Shala (the artificial Lake Koka is not included) have long been known to geologists as the Galla lakes and I will refer to them by that name even though the people of the region are now known as Oromo. I assure them that "Galla" has no personal connotation. A satellite image of these four lakes is shown in Fig. 17.3. Although so close together, they are surprisingly different from each other. Lakes Zwai and Abiata occupy very shallow depressions in the rift floor and have maximum depths of only about 7 m. Both are suffering as a result of human activities, as well as possibly climatic change. Water from Lake Zwai is being used for irrigation of the multitude of flower farms that have sprung up in that part of the rift, and that from Lake Abiata is being channelled and evaporated in order to obtain soda ash (sodium carbonate) for use in glass- making, soaps, detergents and a variety of other products. The ancient shorelines of Lake Abiata, clearly visible on satellite imagery (Fig. 17.3), attest to the rate at which it is disappearing. Lake Langano occupies a faulted basin and has a maximum depth of 48 m (PHOTO 17.6). Lake Shala is the deepest of the four, with its floor plunging to a maximum depth of 257 m. This is because the eastern part of it, as noted above, occupies a caldera, the result of what must have been an unusually dramatic explosion and collapse since all that remains of the original volcano is a low, narrow rim of pumice and ignimbrite.

Even the water in each lake is different. There is good reason why Langano has become surrounded by lodges and resorts, popular destinations for visitors to enjoy swimming and water sports. The lake water, although rather an uninviting brownish colour, is reasonably fresh and is free of bilharzia. There are no crocodiles, and

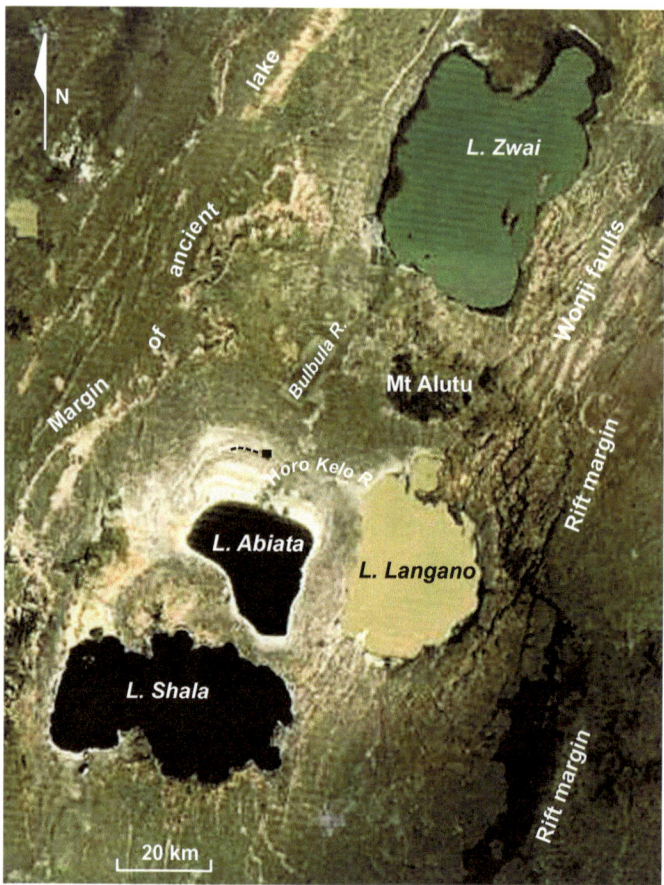

Fig. 17.3 Satellite image showing the "Galla" lakes. Note the previous shorelines of Lake Abiata, the western shoreline of the ancient mega-lake and faults of the Wonji Fault Belt. Although the image does not show true colours, it is clear that those of the four lakes are different. The soda extraction plant at Lake Abiata is shown as a small *black square* and *dotted line* just north of the lake. Image from Google Earth

hippos keep a safe distance away on the northeastern side of the lake. The reason for the fresh water, and for its brownish colour, is that it is close to the eastern margin of the rift and is fed by rivers coming from the Southeastern Highlands. As well as fresh water, these rivers carry particles of red-brown, iron-rich soil which are too

PHOTO 17.6 Fault in ignimbrite forming the southwestern boundary of Lake Langano. Note the brownish colour of the water. A close-up view of the ignimbrite is shown in PHOTO 17.2 (2008)

fine to settle to the lake floor and remain as a brown suspension in the water. Lake Abiata, although only a couple of kilometres from Langano and fed with water from it via the Horo Kelo River, is clear and blue. This is because Lake Abiata has no outlet, so that the water in it has become rich in dissolved soda which coagulates and precipitates the suspended material. This forms a slurry on the lake floor but leaves the water clean and clear. The even deeper blue colour of Lake Shala is due to minute globules of silica suspended in the water, and to its great depth. It is fed by rivers coming from the highlands to the east and west, and by hot and cold springs. Although there is no river connecting it to the other lakes, water trickles through from them by underground seepage. Having no outlet, Shala thus acts as a sump for the water of the entire Galla Lakes region and is very saline (soda-rich). The numerous hotsprings which surround the lake, rising through faults associated with the caldera collapse, also contribute to its high salinity.

These four lakes are remnants of a single huge lake that existed fifty thousand years ago. Prior to this it may have been even bigger, perhaps covering the whole rift floor from Awash to Lake Chamo. The western shoreline of the ancient lake

PHOTO 17.7 Lake sediments exposed in the gorge of the Bulbula River. The *bright white* material is diatomite, a sediment consisting of the microscopic siliceous shells of single-celled algae called diatoms. The *grey layers* consist mainly of volcanic ash and ancient soils (2008)

can be deciphered on Fig. 17.3. The layers of greyish-white sediment, ubiquitous over the rift floor (PHOTO 17.7), were deposited on the floor of this lake. They consist mainly of volcanic ash, fine silty material and diatomite, an accumulation of the microscopic siliceous shells of single-celled algae called diatoms (PHOTO 17.8). These sediments tell a story of changing climates and of how one big lake dwindled into four small ones. This story is illustrated in Fig. 17.4.

Lake Awasa, to the south, is not connected to the Galla Lakes. It occupies a separate basin, on the other side of a watershed which forms quite a prominent rise on the rift floor. Like the northern lakes, Awasa used to be much larger. Ancient Lake Awasa, of which the present-day lake and a small marshy neighbour called Lake Shalo are remnants, occupied a large and now very eroded caldera known as Corbetti caldera. This is itself part of a group of at least two interlocking calderas—with Chabbi volcano sitting on the join, and Urji sitting in the middle of the smaller caldera.

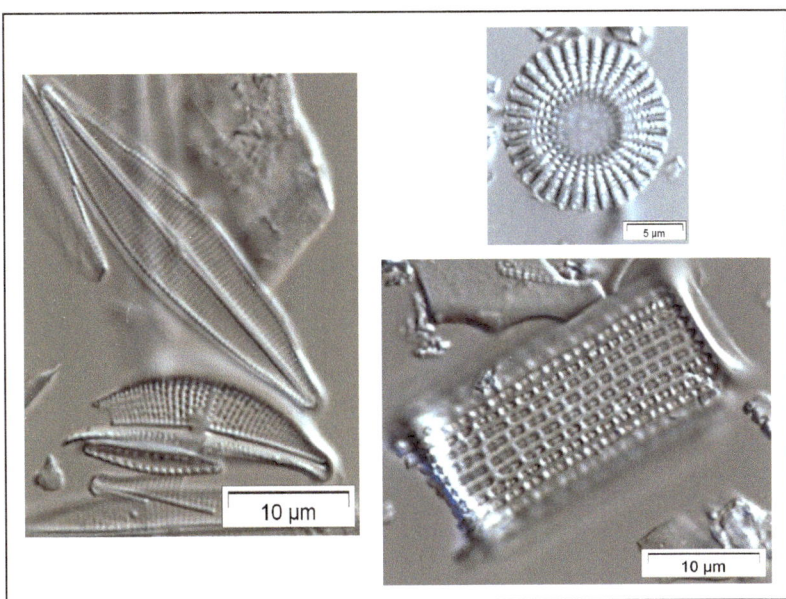

PHOTO 17.8 Diatoms from the Bulbula River gorge, magnified 1000 times (1 μm, or micron, is equal to 1 millionth (10^{-6}) of a metre). These tiny siliceous skeletons accumulate to form diatomite, the soft, white, powdery sedimentary rock shown in PHOTO 17.7. (2011) Photo courtesy Deborah Haynes

A similar story attaches to Lakes Abaya and Chamo further south. Lake sediments on the surrounding rift floor attest to their having been much larger in the past. The two lakes themselves are separated by a ridge of basalt known as Tosa Sucha, or the "Bridge of God",[2] which formed between 1.34 and 0.68 million years ago (PHOTO 17.9).[3] Currently, the only connection between the lakes is a small channel which flows only when their water levels are at their highest. Tosa Sucha is now part of the Nech Sar ("white grass") National Park, best known for its wildlife and particularly its population of zebras (PHOTO 17.10). Lake Abaya, like Langano, is brownish in colour and for the same reason, whereas the water of Lake Chamo is blue.

---

[2]In the local Dorze language *Tosa* means "God" and *Sucha* means "stone".

[3]It is not certain whether the basalt flow caused the separation of the lakes, or whether they were already separated when the basalt erupted.

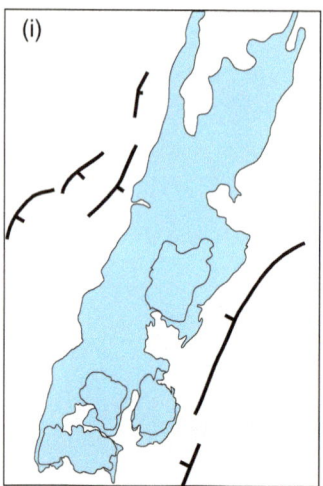

(i) 40,000 - 25,000 years ago. A single lake extended from north of the present position of Mojo to the southern shore of Lake Shala. About 25,000 years ago it began to dry up as the climate became drier.

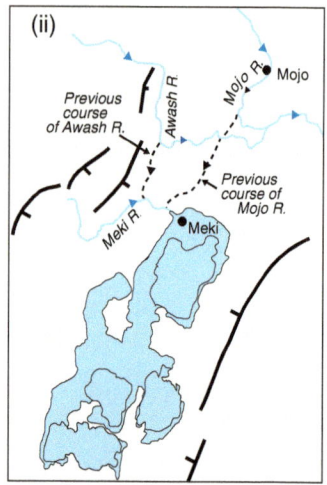

(ii) 25,000 to10,000 years ago. The lake was shrinking due to the drying climate and to earth movements which caused the Awash and Mojo Rivers, the main feeders of the lake, to change course.

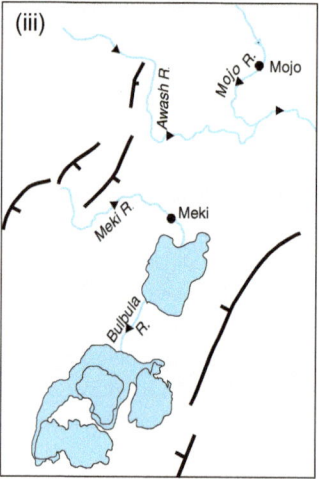

(iii) 10,000 to 3,000 years ago. Rainfall continued to decrease. Despite making brief recoveries from time to time ithe lake continued to shrink.The northernmost part became isolated as Lake Zwai.

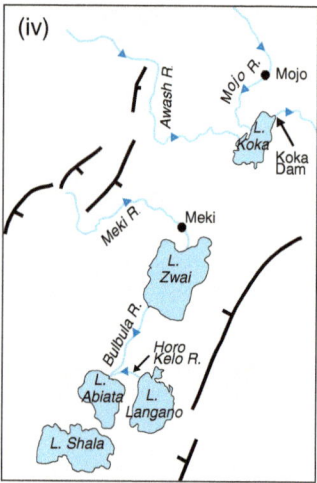

(iv) 3,000 years ago to the present. The remainder of the lake separated into three. In 1960 a new, artificial lake, Koka, was formed by damming the Awash River.

Fig. 17.4 The story of the Galla Lakes: how a large lake became four small ones. Modified from Sagri et al. (2008)

PHOTO 17.9 View southward over the southern end of Lake Abaya (to the *left*) and Lake Chamo (in the distance to the *right*). In the centre, separating the two lakes, is Tosa Sucha or the "Bridge of God". The Amaro Mountains can be seen to the *left* of the lakes, and behind these, topped by clouds, the eastern escarpment of the rift (2014)

17.4   The Wonji Fault Belt

The striking flatness of the rift floor is therefore a result of much of its being an ancient lake bed. It is not, however, entirely flat. We have already referred to the line of volcanoes and the numerous cinder cones which provide quite dramatic relief. In addition, it is dissected by hundreds of young, fresh faults producing escarpments which can run for a few tens of metres to several kilometres. The escarpments may be up to 100 m high which, although small in comparison to the rift margin faults, renders them a conspicuous feature. The faults are not evenly distributed over the rift floor, but tend to be concentrated in bands. The most prominent of these bands, in this part of the rift, runs close to its eastern side, and is known as the Wonji Fault Belt after a village of that name near Gedamsa caldera (PHOTO 17.11). The Wonji Fault Belt can be traced from the region of Lake Abaya for the entire length of the Main Ethiopian Rift and into southern Afar.

PHOTO 17.10 Hills of recent basalt, on the "Bridge of God" in Nech Sar National Park. In the foreground is the *white grass* for which the park is named, and for scale is a group of the zebras for which it is best known (2014)

Geodetic measurements (precise measurements of the earth's surface using, these days, satellite data and global positioning systems) indicate that in this region the rift is widening at a rate of 0.3–0.5 cm per year, and that most of the widening is occurring across the Wonji Fault Belt. In addition, most of the earthquake activity in the rift takes place along the Wonji Fault Belt rather than along the main rift margin faults. These move much less frequently, and are becoming worn and eroded whereas the Wonji faults are sharp and fresh. The rifting activity, which began with the faulting of the rift margins and great eruptions of ignimbrite along them, has moved inwards to the rift floor. This gradual concentration of rifting activity into a confined axis represents an important stage in the break-up of a continent. There will be more about this when we come to Chap. 25.

PHOTO 17.11 Faults of the Wonji Fault Belt, east of Lake Zwai (2008)

17.5 The Amaro Horst

Southwards from the point at which Lake Abaya narrows, about halfway along its length, the rift undergoes a number of changes. Firstly, the "typical" rift features: the ignimbrites, the volcanoes and the Wonji Faults, disappear. Then, southeast of Lake Abaya, we meet something we have not seen since Chap. 15: Precambrian rocks. They form a mountain range known locally as the Amaro Mountains, and geologically as the Amaro Horst. A horst is a block of land which has moved upwards between two faults, as shown in Fig. 17.5. The Precambrian rocks of the

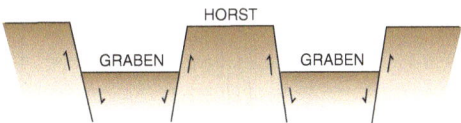

Fig. 17.5 Graben and horsts

Amaro Horst are the highest found anywhere in Ethiopia, over 2600 m—so have undergone a considerable amount of uplift here! They are partially capped by volcanic rocks, corresponding to flows of the Trap Series.

The horst is a substantial feature, 90 km long, 25 wide and rising 1500 m above the rift floor. The rift splits into two sections around it: the Ganjuli Graben and the Galana Graben. Graben is a German word[4] meaning "ditch" or "trench" and in geology it is used, like rift valley, to describe a strip of land that has subsided between two opposing faults. "Graben", however, generally refers to a smaller-scale feature whereas a rift valley is something on a continental scale. A graben is the opposite of a horst, as shown in Fig. 17.5. Lake Chamo and the southern half of Lake Abaya occupy the Ganjuli Graben.

The Amaro Horst and its neighbouring graben are indications that the Main Ethiopian Rift is beginning to lose its identity here, and to break up. South of Lake Chamo it becomes hard to recognise a rift at all. The rift margins become indistinct and the region is broken up into a broad zone of faults, horsts and graben known as the Broadly Rifted Zone. This is a transition region between the Main Ethiopian Rift and the Kenyan Rift, and we will visit it in the next chapter.

---

[4]Since "graben" is a German word, its plural form "graben", is frequently, and more correctly used rather than "grabens". However, although "horst" is also a German word (meaning "heap"), the plural "horsts" is generally used. Not logical, but there it is!

South of Lake Chamo: A Transition Region

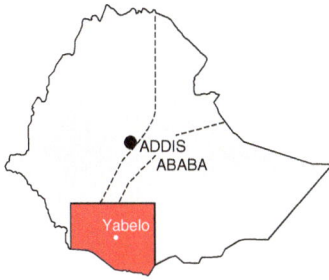

Although southern Ethiopia is most often visited for its exotic peoples and abundant and colourful bird life, its geology is as interesting, varied and complex as the people who inhabit it.

South of Lake Chamo, the scenery undergoes a marked change from that to the north. There is no more flat rift floor, nor clearly defined rift margins, and the dark volcanic rocks of the highlands bordering the rift give way to pink and white metamorphic and igneous rocks of the Precambrian basement. The Rift Valley, such a distinct feature further north, breaks up into a broad region of faults, largely in the form of short and discontinuous horsts (uplifted blocks) and graben (downfaulted blocks). For this reason it is referred to as the Broadly Rifted Zone. To the south, near the Kenyan border, the rift proper recommences where the Omo river flows into Lake Turkana, and from there it continues southwards through Kenya and Tanzania.

The geology of this region is shown in Fig. 18.1. It is a mix of Precambrian rocks, sandy plains, some very recent volcanic features and the southern limit of the Trap Series volcanics. There are also many faults, but only the more prominent of these are shown to avoid cluttering the figure too much. The Precambrian rocks

© Springer International Publishing Switzerland 2016
F.M. Williams, *Understanding Ethiopia*, GeoGuide,
DOI 10.1007/978-3-319-02180-5_18

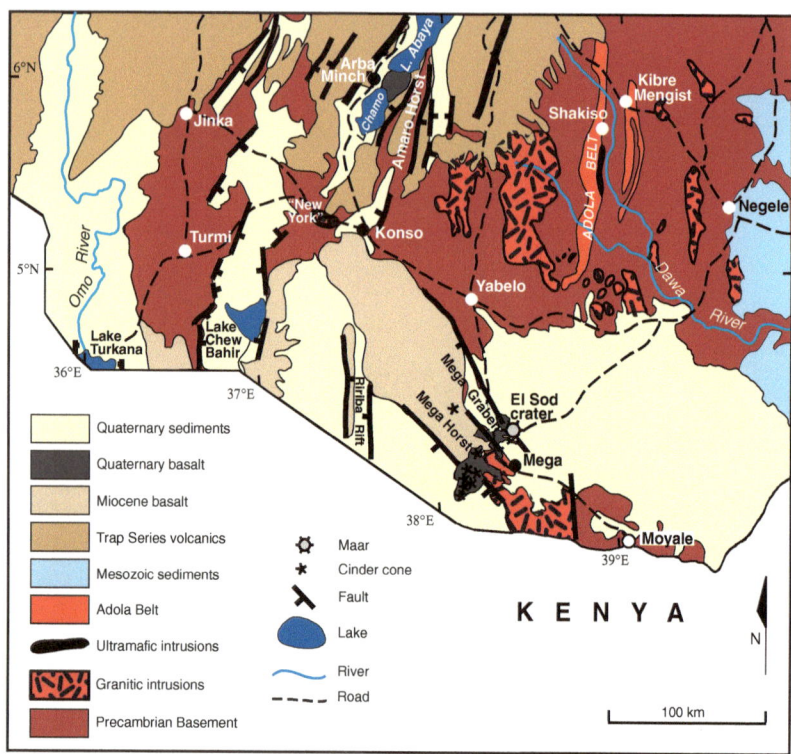

Fig. 18.1 Geological map of southern Ethiopia. Modified from Geological map of Ethiopia 1:2,000,000 2nd Ed. (1996) and Geological map of Yabelo 1:250,000, Geological Survey of Ethiopia

are rather different to the ones we encountered in western Ethiopia. They are almost entirely gneisses and granites belonging to the Mozambique Belt, rather than the metamorphosed volcanic rocks of the Arabian-Nubian Shield that we encountered in the west. The exception is the strip marked in orange-red on Fig. 18.1. This is the Adola Belt, a narrow band of Arabian-Nubian Shield rocks

PHOTO 18.1 Granite inselberg near Turmi (2010). Photo courtesy Emile Farhi

protruding into this region. We will look at the Adola Belt later in this chapter since it contains some unusual and interesting rocks, and is very important to Ethiopia's economy.

Much of the attractive scenery of this southern part of Ethiopia is due to the way that the granites and gneisses weather into small, rounded hills called inselbergs (PHOTO 18.1). These may weather further, by the process which we saw in Chap. 15, to become a pile of boulders known as a tor (PHOTO 18.2). In between the hills and rocky outcrops the granites and gneisses have broken down completely to a coarse sandy soil, from which termites construct pillar-like mounds (PHOTO 18.3). Occasionally gemstones such as sapphire and tourmaline weather out of the Precambrian rocks and may, if you are very lucky and look very hard, be found among the coarse sand grains in dry river beds (PHOTO 18.4).

PHOTO 18.2   Granite-gneiss tor near Yabelo (2014)

18.1   "New York"

West of the town of Konso is a feature known to tourists as "New York", for reasons which become obvious when you see it. A surreal landscape of red towers and pinnacles, supposedly reminiscent of the skyscrapers of that famous city, extends across the floor of a broad canyon (PHOTO 18.5). The rock of which these are formed is a gneiss or granite which has weathered to a soft, almost sandy material. This is probably largely due to the climate of this part of the country. The rock, already weakened by groundwater seeping through it, is subjected to the

PHOTO 18.3 Termite mound near Mega, built from the loose, granitic soil. Tawny eagle for scale (2014)

alternating temperatures of the warm days and cold nights which, repeated thousands of times, cause the rock to expand and contract and gradually break up into its individual grains. This loose material is easily washed away by rainwater and by streams flowing across it. The pillars are what remain after everything between them has gone. Originally there may have been a capping of hard basalt which protected their tops from erosion—rather like the mesas and pillars that we encountered in Tigray in Chap. 11—but if so this capping has now gone. Features like this are known in America as hoodoos; elsewhere and rather more evocatively as fairy chimneys.

If you visit "New York" you may be besieged by children trying to sell you pieces of an attractive blue-green mineral called amazonite. This is a form of feldspar, whose bluish colour is thought to be due to small amounts of lead impurities in the crystal structure. It probably formed in pegmatite veins within the gneiss or granite. It is not a very common mineral, and some of the best specimens of it in the world come from this locality.

PHOTO 18.4   Sapphire in river gravels, Yabelo area. Photo courtesy Kassaye Yohannes

PHOTO 18.5   "New York": spires and pinnacles formed by the erosion of weathered, crumbly granite gneiss (2014)

18.2   Stone Age Peoples

The faulted basins containing Lake Turkana and Lake Chew Bahir, and a smaller area north of Konso, contain the sediments of ancient rivers and lakes. These sediments consist, like those of the Rift Valley lakes further north, of silt, clay and volcanic ash. They are between about 1 and 2.5 million years old, and have proved to be rich hunting grounds for palaeo-anthropologists (specialists in the study of our early human ancestors) as they contain much evidence that the ancient river banks and lake shores were occupied by Stone Age peoples. Stone tools and fossil remains of *Homo erectus* ("upright man"—one of our direct ancestors) and *Australipithecus robustus* (his more ape-like cousin) have been found, together with the remains of many of the mammals which early man may have hunted.

18.3   El Sod: An Explosion Crater or Maar

Near to the town of Mega, surrounded by a wide expanse of mainly Precambrian territory, it seems quite surprising to come upon what is clearly a recent volcanic crater, a sudden and unexpected hole in the ground. This is El Sod—a crater about 1.8 km in diameter and 340 m deep with a very low rim consisting of layers of ash, tuff and rock fragments (PHOTO 18.6). At the bottom is a small black lake saturated with salt, locally known as black salt, which is mined by the local people and carried out of the crater on donkeys. The salt itself is not actually black; the colour is due to black organic matter which is mixed with it.

El Sod is an example of what is known as a maar, and we shall meet more of these features when we look at the rift margins in Chap. 20. A maar, also known as an explosion crater, is produced by a steam explosion. Hot magma rising to the surface encounters water, either in the form of a lake or simply the water held in the ground. This instantly evaporates into steam, which bursts through to the surface, creating a deep hole and throwing out rock fragments, ash and lava droplets which pile up around the vent and are referred to as ejecta. The eruption occurs as a series of pulses, maybe several hundred, each of which throws out a pile of debris so that layer upon layer of ejecta form around the crater rim. El Sod is a slightly unusual maar in that the amount of rim ejecta appears very small compared to the size of the crater. The reason is that most of the ejecta are not around the rim. The explosions were so violent that they were flung a long way and are spread over a large area. Deposits of them can be seen at least 10 km away, near the main Addis Ababa to Moyale road.

PHOTO 18.6 El Sod maar, or explosion crater, near Mega. The dividing line between the gneiss cliffs which form the crater walls, and the overlying rim ejecta, can be seen quite clearly near the top of the crater (2014)

Another interesting feature of El Sod is that the rim ejecta contain, as well as blocks and fragments of the gneiss through which the eruption occurred and which forms the steep crater walls, nodules of a green mineral called olivine (PHOTO 18.7). Olivine is a constituent of the earth's mantle and the presence of the nodules indicates that the eruption must have started very deep down, carrying up material from the mantle itself, from some 50 to 70 km below the earth's surface.

Why is this crater here, in the middle of Precambrian basement rocks? It is not in fact the only volcanic feature in the region. Although the scale of Fig. 18.1 is too small to show them, a number of similar maars are present, as well as clusters of cinder cones and flows of fresh, scoriaceous basalt. This volcanic activity is associated with faulting which took place in Quaternary times and which resulted

PHOTO 18.7 Olivine nodule from the rim deposits of El Sod maar: material from the earth's mantle! (2014)

in a raised strip of land called the Mega Horst, and an adjacent sunken strip, the Mega Graben. They are a pair of the many horsts and graben that occur right across this Broadly Rifted Zone, though not all were accompanied by volcanic activity like the Mega ones.

18.4   The Adola Belt: Gold and Tantalum

The Adola Belt (named after Adola, the Oromo name for the town currently marked on most maps as Kibre Mengist) is a southern extension of Arabian-Nubian Shield rocks. It is shown on Fig. 18.1 as a narrow, orange-red strip. During the East African Orogeny this narrow zone of rocks had been compressed, squeezed into folds, squeezed and folded at least one more time, and finally sheared—stretched sideways so that layers of rock were dragged alongside each other. The original rocks were changed beyond recognition, as heat and

PHOTO 18.8 Panning for gold in the gravels of the Dawa River (2010). Photo courtesy
Emile Farhi

pressure caused new minerals to form and generated watery fluids in which rare
elements became concentrated. These fluids seeped along the cracks created by the
shearing process and eventually solidified as veins of quartz or coarse-grained
pegmatite. Contained in these veins are rare elements such as beryllium, lithium,
tantalum and gold.

Gold does not combine with any other element; it stays as just gold. Large
lumps of gold are known as nuggets, but in this region nuggets are rare and the
gold occurs mainly as scattered flakes and small fragments. It is extracted in two
ways. At the Lega Dembi Mine near Shakiso it is obtained by digging out,
crushing and processing the gold-bearing rock itself, known as the primary gold
deposit. This requires a substantial and expensive operation. The second method is
to let nature do some of the work, by using rock which has already been weakened
and broken down by weathering. Rivers carry away the weathered material, and the
gold flakes and fragments are dispersed in the sand and gravel of the river beds.
This type of deposit is known as a placer deposit. Extraction of the gold takes

PHOTO 18.9 Spodumene (a lithium aluminium silicate mineral) in pegmatite at Kenticha tantalum mine. Spodumene forms elongated, tabular grains, which here have been weathered to a *brownish colour* (2014)

advantage of the fact that it is heavy. A common technique is panning, in which a shallow pan containing gravel and water is swirled gently so that centrifugal force sends the lighter gravels to the edge of the pan, leaving (if you are lucky!) a concentration of gold in the centre (PHOTO 18.8). Every river bed in the Adola region is dotted with gold panners, working as individuals or in groups and hoping to make their fortune, or at least eke out a living, from gold. A variation of this method is employed on a large commercial scale at the Adola gold mine near Shakiso, where bulldozers dig out gravel that has been deposited in ancient river beds. This is then treated by mechanical procedures before finally being panned by hand.

Not far from the town of Kibre Mengist (Adola) is the Kenticha Tantalum Mine. The rare element tantalum is used in electronic components and occurs naturally in the form of the mineral tantalite, an oxide of tantalum, iron, manganese and niobium. At Kenticha tantalite occurs, like gold, in pegmatites, together with quartz

PHOTO 18.10 Hand-sorting tantalite (the *dark-coloured* grains) from quartz at Kenticha tantalum mine (2014)

and a tabular mineral called spodumene (PHOTO 18.9). The pegmatite is very weathered and breaks up easily. The tantalite is extracted simply by flushing the weathered pegmatite with water and finally picking out the tantalite grains by hand (PHOTO 18.10). The tantalum itself can then be obtained using chemical procedures. It has become a very important material in the 21st century as it is a component of many of the devices on which we have become dependent, such as mobile phones, laptop computers and DVD players.

The Rift Valley Northwards: Volcanoes, Fissures and Fresh Lava Flows

19

A short distance north of the town of Nazret (Adama), the scenery within the Main Ethiopian Rift changes. It becomes harsher and more desolate. The lake sediments, such a widespread feature of the rift floor further south, disappear, to be replaced by a scatter of black basalt boulders and cindery blocks. At first these are thinly spread, but become thicker and more concentrated as one proceeds northwards. Signs of cultivation give way to thorny scrub. Cinder cones are fresher and more abundant (PHOTO 19.1). The climate becomes markedly warmer between Nazret, at an elevation of 1700 m, and Awash at 950 m, and the people too are different. Here are the slender Karrayu in their long sarongs, the men often with a stick or a rifle held across their shoulders, and perhaps driving a camel or two.

This section of the rift is shown in Fig. 19.1 as a digital elevation model, and Fig. 19.2 shows a geological sketch map. Much of the rift floor here is, like that further south, underlain by ignimbrite, but along its central region this is covered by flows of fresh basalt which have erupted through fissures, some of which can still be seen. The uppermost layers of these basalt flows have weathered to give the scatter of boulders mentioned above. Lake sediments are few and patchy—but there is plenty of dust, as anyone travelling in the region will quickly discover.

© Springer International Publishing Switzerland 2016
F.M. Williams, *Understanding Ethiopia*, GeoGuide,
DOI 10.1007/978-3-319-02180-5_19

PHOTO 19.1 A fresh cinder cone, near to K'one volcano in the northern Main Ethiopian Rift (2010)

Faults of the Wonji Fault Belt form prominent scarps, and the line of volcanoes which we encountered further south continues. In fact both the Wonji Fault Belt and the line of volcanoes extend well into southern Afar, but we will come to that in Chap. 23.

19.1 Volcanoes of the Northern Rift

The volcanoes of the northern rift: Boseti Guda, Boseti Bericha, K'one and Fantale are, like those further south, in a dormant state, and are formed of silica-rich rocks such as trachyte, rhyolite, pitchstone and pumice. Although fairly similar to each other in composition, they are quite different in structure. Boseti Guda and Boseti Bericha are prominent cones with summit craters; Fantale has a small but spectacular caldera (for those sufficiently energetic to climb the mountain!) and K'one is almost entirely caldera. All are associated with flows of fresh basalt, which have

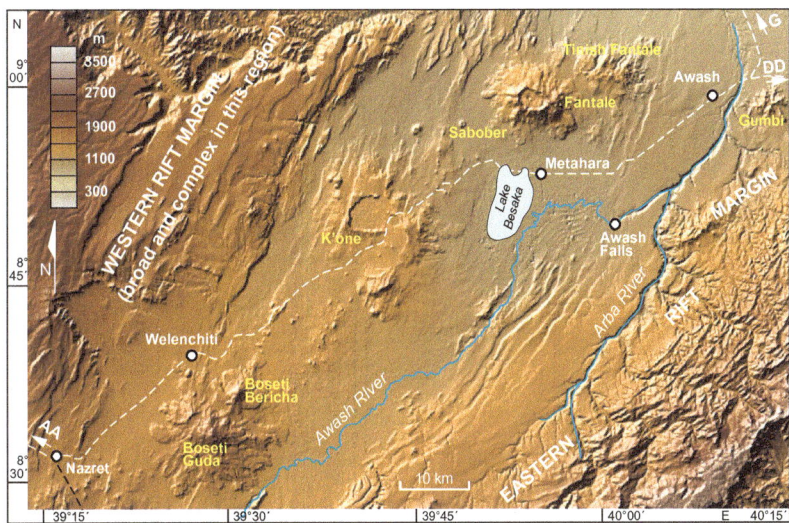

Fig. 19.1 Digital elevation model of the northern section of the Main Ethiopian Rift. The *white dashed line* indicates the main road from Addis Ababa (AA) to Dire Dawa (DD) and Gewani (G). DEM from GeoMapApp

not come from the volcanoes themselves but from fissures on which they are located. We will look at just two of these volcanoes, K'one and Fantale, in more detail as they are readily accessible and exceptionally interesting.

19.2 K'one Volcano

A traveller from Nazret to Awash can hardly miss K'one volcano as the main road passes right through it. However, it does not conform to the conventional picture of a volcano. It is not a mountain—but a complex of calderas—broad circular depressions surrounded by low rims of pumice and ignimbrite. Like Shala further south, the top of the original volcano (or in this case volcanoes) was blown off in a violent eruption, or series of eruptions, and whatever remained sank back as the magma chamber below emptied. There are possibly as many as seven calderas in the K'one complex, interlocking with each other. The two most conspicuous ones are located to the right and the left of the Nazret-Awash road. Unlike Shala, whose caldera is filled with a lake, these are filled with blocky basalt lava. Although this is

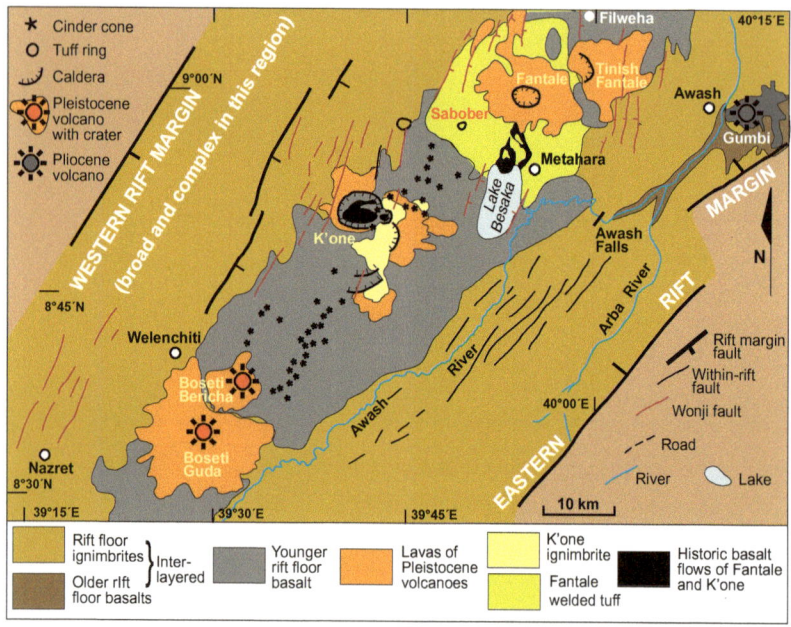

Fig. 19.2 Geological map of the northern section of the Main Ethiopian Rift. Modified from Geological map of Ethiopia 1:2,000,000 2nd Ed. (1996), Geological Survey of Ethiopia

rapidly becoming vegetated, it still appears quite black and fresh as do the cinder cones located along the junction of the calderas. The basalt appears to have flowed from a fissure system running along this junction. These two calderas are the youngest in the complex; the older ones are not so easy to make out as ignimbrite from later eruptions has poured over them, filling and partially obliterating them.

There is an interesting account given by Major W. Cornwallis Harris in his book "The Highlands of Ethiopia (Volume 3)". Major Harris was British ambassador to the then kingdom of Shoa from 1841 to 1844, and in 1842 he made an excursion through this region, passing by what he describes as the "*great crater of Winzegoor*". From the description of his route and of the area this corresponds to K'one. He states that the cinder cones were "*thrown up during an eruption some 30 years previously*". Unfortunately he does not give the source of this information, but the fresh appearance of the cones and of the basalt filling the calderas is certainly consistent with such a recent date.

It is not known how high the original K'one volcanoes were, as they have been almost entirely destroyed in the explosion, but judging by the size of their calderas they would have been big!

19.3 Fantale Volcano and Awash National Park

Situated at the apex of the Main Ethiopian Rift, where it opens out into Afar, is the magnificent volcano Fantale (PHOTO 19.2). Fantale is also the centrepiece of the Awash National Park, which is itself a treasure trove of interesting geological features (Fig. 19.3). The volcano, which rises about 600 m above the surrounding plain, has a summit caldera measuring about 2.5 by 3.5 km. The formation of the caldera was associated with the explosive eruption of nuée ardente which rolled down the mountainside and solidified to a particularly beautiful welded tuff

PHOTO 19.2 Fantale volcano, the fresh lava flow and Lake Besaka. A small tuff ring known as "Tinish Sabober", with a cone in its centre, can be seen just in front of Fantale. The point from which this photo was taken is now submerged beneath the waters of Lake Besaka (2006)

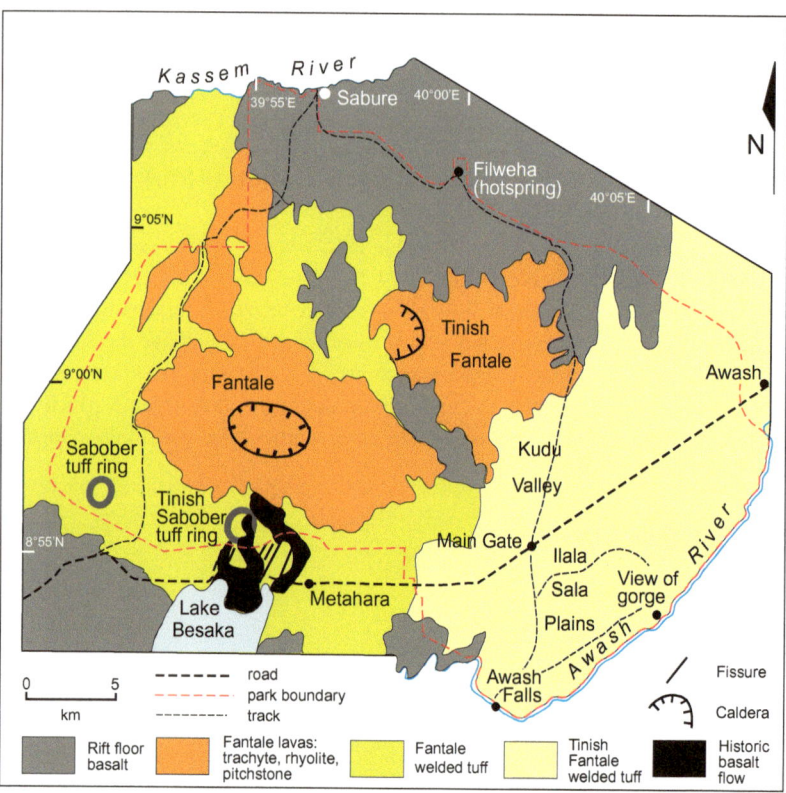

Fig. 19.3 Geological sketch map of Awash National Park. The location of the railway line is not shown, as it was still under construction at the time of writing. Individual volcanic blisters which are scattered over the welded tuff plain are not shown as they are too numerous

(PHOTO 19.3). Welded tuff (as you will remember from Chap. 17) is an ignimbrite in which the heat has been sufficient to fuse together the ejected fragments. Sometimes, as in the Fantale welded tuff, blobs of lava have been drawn out to solidify as black glassy streaks called fiamme (an Italian word meaning "flames"). This, together with older welded tuffs and ignimbrites of the rift floor, with some interbedded layers of basalt, forms the flat plain upon which the volcano stands and which provides the wide grasslands which are home to oryx, kudu, warthog and other, unfortunately diminishing, wildlife of the National Park.

PHOTO 19.3 Close-up of the Fantale welded tuff. The *grey* streaks are stretched fragments of pumice; the *black* streaks are fiamme, liquid lava droplets that have solidified as a black glass like obsidian. *10 cent coin* for scale (2010)

As well as being home to these animals, the welded tuff plain contains a number of unusual geological features. Small but conspicuous hummocks, a few metres to tens of metres in diameter, dot its surface (PHOTO 19.4). Some of these have collapsed (PHOTO 19.5), showing that they are hollow inside. They are actually enormous gas bubbles, formed by gas trapped inside the welded tuff when it solidified. They have been termed volcanic blisters since they resemble the blisters which form on your skin by liquid trapped beneath, or those which form in poorly applied paintwork. To my knowledge, this is the only place in the world where such features have been found in welded tuff, though they are not uncommon in basalt.

Just south of the volcano, the welded tuff plain is cut by a series of at least six open fissures, up to more than a metre in width, 22 m in depth and 2.5 km in length

PHOTO 19.4 Unbroken blister—a huge gas bubble—in the Fantale welded tuff (2014)

(PHOTO 19.6). They are tensional fissures, a visual reminder that the rift is stretching, fracturing and being pulled apart as the Somali and African Plates separate, which in this region they are doing at a rate of about 5 mm a year.

Also to the south of Fantale is a field of fresh, black, blocky basalt, much like that which occupies the calderas at K'one (see PHOTO 19.2). As noted above, the lava does not come from the volcano itself but from a fissure system on which it is located. The date at which this flow erupted is often quoted as 1810 or 1830 AD, again on the basis of an observation by Major W. Cornwallis Harris. Although it is not entirely clear which volcanic feature he was referring to, the fresh appearance of the basalt is, as in the case of K'one, consistent with a very recent eruption. It is interesting to note that Major Harris reports abundant wildlife in the region. He observes that. "*long-haired oryx, with great herds of buffalo, grazed around every pool*" and "*formidable troops of (lions) which, roaming almost unmolested,*

PHOTO 19.5 A broken blister, showing that it is hollow inside. Mt Fantale is in the background (2012)

commit great havoc among the cattle." Sadly, most of these animals can no longer be seen in Awash National Park—with the exception of the cattle!

Prominent on the welded tuff plain to the southwest of Fantale is a distinctive hill known as Sabober (PHOTO 19.7). Sabober is a tuff ring which, as its name implies, is a ring of tuff (volcanic fragments) surrounding a central crater. A tuff ring is similar to a maar, but does not have such a deep crater and layering of the rim deposits is less apparent. As in the case of a maar, the eruption has occurred through water causing a steam explosion, but the source of the eruption is not as deep. There is another tuff ring, known as Tinish ("little") Sabober, directly to the

PHOTO 19.6 One of the several open fissures in the Fantale welded tuff (2012)

south of Fantale (see PHOTO 19.2). This has a small central cone, and at first sight it appears to be the source of the fresh lava flow. This however is not the case; it just happens to be located on the same fissure system, as is Fantale itself. To the northeast of Fantale is a similar, but older, flatter and more eroded volcano known as Tinish Fantale, whose caldera has been partly obliterated by the welded tuff from its younger neighbour.

Other attractive features in the Awash National Park are the picturesque hot-springs known as Filweha (which simply means "hot water"—and beware, it is!) (PHOTO 19.8), the Awash Falls where the Awash River plunges over a barrier of basalt (PHOTO 19.9), and the Awash gorge downstream of the falls where the river has cut a spectacular section through the layers of ignimbrite and basalt that form the rift floor here (PHOTO 19.10). The falls are a smaller version of, and formed by a similar process to, the Tis Isat Falls of the Blue Nile.

PHOTO 19.7   Sabober tuff ring, looking toward the northwest (2014). Image from Google Earth

19.4   The Puzzle of Lake Besaka

The southern part of the field of fresh basalt, south of Fantale, is submerged beneath the waters of Lake Besaka. In the 1970s this was a small lake, about 1.5 km wide and not much longer. At the time of writing (2015) it is over 5 km wide and 10 km long (Fig. 19.4). The main road and the railway line linking Addis Ababa and Dire Dawa, which were originally constructed well to the north of the lake, have been inundated and had to be raised numerous times over the years in order to stay above water. By the year 2013 the rising water had become too much (PHOTO 19.11), and the road and railway line have since been diverted to the north through the recent basalt flow, which unfortunately has been severely degraded in the process.

Various theories have been put forward to explain this extraordinary increase in the size of the lake, including increased activity of the hot and cold springs which feed it, movement of faults, and even leakage of water from the Koka dam 90 km

PHOTO 19.8   Pool fed by the hotsprings at Filweha, northeast of Mt Fantale (2010)

to the southwest, seeping along fault lines. The most plausible theory is that irrigation procedures in the fruit plantation south of the lake have caused the level of the groundwater, and hence of the lake, to rise. Recently the problem may have solved itself, as an overflow channel has cut its way through from the lake to the Awash River, thus maintaining the current lake level.

19.5   The Wonji Fault Belt in the Northern Main Ethiopian Rift

The Wonji Fault Belt, which we encountered in Chap. 17, is also a prominent feature of the northern section of the rift. However, because the rift here has changed its orientation from NNE-SSW to NE-SW, the faults are no longer parallel to the rift itself and the belt as a whole gradually approaches to the western rift margin (see Fig. 19.1). In fact, north of Nazret the Wonji faults run right up against the rift margin and become indistinguishable from the margin faults themselves. In

PHOTO 19.9 Awash Falls, where the Awash River drops over a ledge of hard basalt (2014)

order to keep within the rift, the whole belt has to shift eastwards. North of K'one it shifts eastward again, and continues to make eastward shifts as it extends into southern Afar. It is thus made up of short segments, each displaced eastward from the previous one in what is described as an en échelon (from the French word *échelon* meaning "rung of a ladder") progression.

The reason for this en échelon pattern, and why the Wonji faults are not parallel to the rift margins, has been much debated. One idea is that the direction of extension (pulling) has changed over time—from NE-SW during the Pliocene epoch when the rift margin faults were forming, to an E-W direction as the faulting activity moved from the rift margins to the rift floor. It is likely that, whatever the reason for the discrepancy, the individual Wonji faults are aligned with a direction of structural weakness, or "grain", in the Precambrian basement, inherited from the events of the East African Orogeny. This "grain" may result from the alignment of fibrous or platey minerals, or may be the result of sideways, shearing movements, and is the natural direction for breaks to occur. It is like trying to tear a piece of material. Whichever way you pull it will shred along its grain even if the tear itself

PHOTO 19.10 The Awash gorge, downstream of the Awash Falls. The river has carved its way downward and headward through layers of basalt and ignimbrite. Columnar jointing can clearly be seen in some of the layers (2012)

Fig. 19.4 Lake Besaka. *On the left* is an aerial photograph of Lake Besaka taken in 1957, and *on the right* a satellite image taken in 2015. The lake has expanded more than 10 times in area. Satellite image modified from Google Earth

PHOTO 19.11   The end of the road! The water of Lake Besaka has risen so high that the road cannot be raised any further, and has had, together with the railway line, to be diverted to the north (2013)

goes the way you have been pulling. You can try this if you have an old sheet—even a tissue may work!

It is notable that the volcanoes Boseti Guda, Boseti Bericha, K'one and Fantale sit upon segments of the Wonji Fault Belt. It seems that molten material beneath the surface is providing a kind of lubricating effect which makes it easier for rocks to slide past each other. This facilitates faulting, and the faults then provide conduits for the magma to rise to the surface and form the volcanoes. Magma, faulting and volcanic activity thus work hand-in-hand in the processes involved in rifting. We will see further examples of this when we move to Afar in Chaps. 21–23, but first we will take a look at the very important transition zone which separates the highlands from the rift and Afar.

The Rift Margins and the Great Western Escarpment

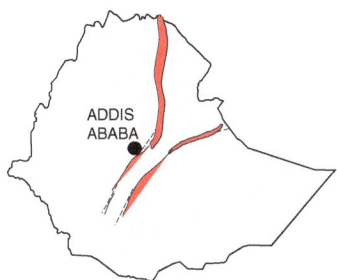

Although the margins of the Rift Valley and Afar are often shown as distinct black lines on geological maps, including those in this book, they are in reality much more complex. On the ground, indeed, they are often difficult to decipher. It is easy to be certain when you are in the highlands, or when you are in the Rift Valley or Afar, but it is less clear just where one ends and the other begins. This is because the rift margins are not formed of single, simple faults but are in most cases broad, complex belts of faults and flexures, extending over widths of up to 80 km. Sometimes the original faults have been obscured by later flows of lava or ignimbrite. Finally, they have been subject to millions of years of erosion by rivers and streams cutting across them, wearing them back and blurring their original features.

The rift margins have different patterns of faults in different places, and sometimes there are no obvious faults at all, just slopes or flexures. Figure 20.1 illustrates some of the profiles that the rift margins assume. It is by no means intended as a classification. Such a thing would be impossible on account of their complexity, the fact that there is every gradation between them even along the same margin, and that there are many sections of margins which would not fit any category. What form they take depends on numerous factors such as the rock type, the presence or absence of old lines of weakness and the nature and direction of the forces involved, but they have one thing in common: they are zones of extension of

© Springer International Publishing Switzerland 2016
F.M. Williams, *Understanding Ethiopia*, GeoGuide,
DOI 10.1007/978-3-319-02180-5_20

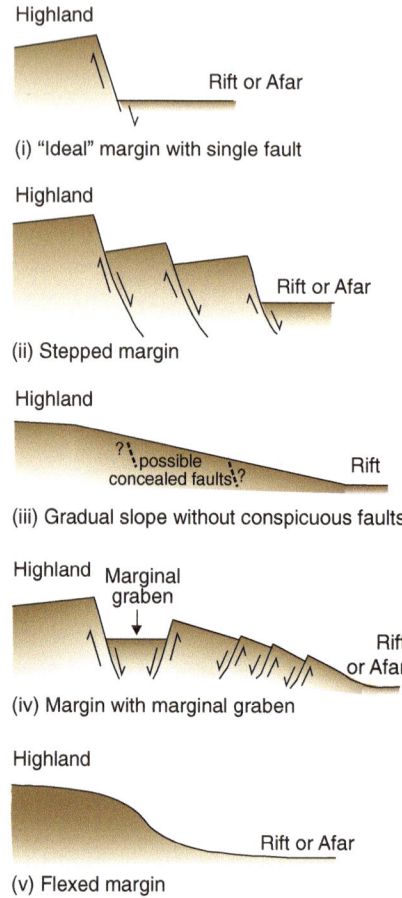

Fig. 20.1 Idealised profiles illustrating some of the forms that rift margins can take. Note that they are not drawn to scale

the earth's crust. In profile (v) the crust has stretched, thinned and sagged. If you imagine the profiles (i), (ii) and (iv) to be put back the way they were before the faulting occurred, they would all be shorter. This is of course because they are being pulled and stretched as the African, Arabian and Somali Plates move apart.

Another common feature of the rift margins is that where faulting is present, as in Fig. 20.1(i), (ii) and (iv), the faulted blocks tilt back toward the highlands. Figure 20.2 explains how this happens. The rim, or shoulder, of the margin is thus

Before faulting this amount of rock
is pressing down on the future
fault face.

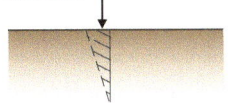

After faulting the rock is no longer
pressing down on the fault face,
so the edge of the faulted block
flips upwards.

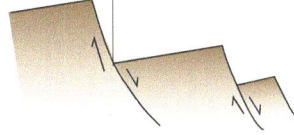

As they get deeper the fault
planes curve upwards slightly,
tilting all the blocks.

Fig. 20.2   Illustrating how a rift margin fault becomes tilted upwards

higher than the adjacent highland. Since the Trap Series volcanics also tend to be
thickest in the region of the rift margins, this results in imposing mountain ranges,
particularly along the western margin of Afar which coincides with the greatest
uplift of the Afro-Arabian Dome. Apart from the great shield volcanoes of the
highlands, such as the Semien and Bale Mountains, Ethiopia's highest mountains
are located atop the rift margins.

20.1   The Margins of the Main Ethiopian Rift

Figure 20.3 shows the margins of the Main Ethiopian Rift, and that of southern
Afar. You can see that the eastern and western margins of the Main Ethiopian Rift
are rather different from each other. For most of its length, from Sire to the latitude
of Lake Chamo, the eastern margin is quite narrow and well defined. In the region
of Sire it consists of two to four main steps, like the profile shown in Fig. 20.1(ii),
each separated by a broad, gently sloping "tread" covered by fertile land and
cultivated fields. East of Lake Langano a group of big volcanoes Chilalo, Kaka,
Badda, Encuolo sit atop it, and the lavas of Mt. Kaka have partly obscured the rift

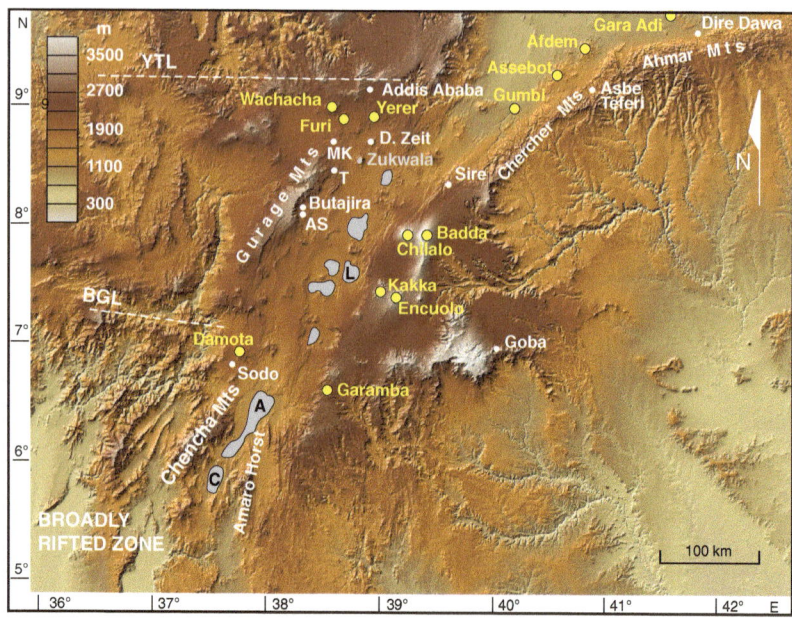

Fig. 20.3 Digital elevation model showing the Ethiopian Rift Valley and its margins, and the southern margin of Afar. To save space the following abbreviations are used: *L* Lake Langano; *A* Lake Abaya; *C* Lake Chamo; *MK* Melka Kunture; *T* Tiya; *AS* Ara Shatan. The *white dashed lines* are the Yerer-Tulu Welel (YTL) and Bonga-Goba (BGL) lineaments that intersect the western Rift margin. DEM from GeoMapApp

margin faults. These volcanoes are around 2 million years old and, as noted in Chap. 15, may be connected with the later stages of the rift margin faulting. South of Mt. Kaka the margin runs in a straight line for almost 200 km, as near as can be to the "ideal" rift margin shown in Fig. 20.1(i). East of Lake Chamo, where the rift splits around the Amaro Horst, both rift and margin disintegrate into the Broadly Rifted Zone of southern Ethiopia.

The western margin of the Main Ethiopian Rift is more complex. It ranges from having no escarpment in the regions of Addis Ababa and Sodo, an arcuate one near Butajira, and a fairly well defined one bordering Lakes Abaya and Chamo which becomes less distinct as the Broadly Rifted Zone is approached. We will look at these sections in a little more detail.

20.2   The Rift Margin in the Addis Ababa Region

In the Addis Ababa region there are no obvious rift margin faults. The descent from Addis Ababa into the Rift Valley is so gradual as to be hardly noticeable. It is like the profile in Fig. 20.1(iii). In fact Addis Ababa could be regarded as being in a kind of embayment of the rift, and the Entoto Mountain range behind the city as being the start of the highlands proper. The complex situation here is partly due to the Yerer-Tulu Welel lineament—the broad E-W trending band of faults and volcanoes which we discussed in Chap. 12—which intersects the rift margin at this point.

A number of interesting features mark the gradual transition between Addis Ababa, and the Rift Valley proper with its flat floor and characteristic beds of lake

PHOTO 20.1 Mt Wachacha, a Pliocene (about 4 million years old) volcano near to Addis Ababa, viewed from the west (2012)

sediments. Firstly, there are three big volcanoes, Furi, Yerer and Wachacha, to the south and southwest of the city (PHOTO 20.1). These volcanoes are formed of trachyte and are Pliocene in age (about 4 million years old), a little older than Mt Chilalo and the volcanoes on the eastern rift margin. Like them, they appear to be connected in some way with the rifting process. A line of cinder cones extends southwestward from Mt Yerer across the mouth of the embayment and may indicate some concealed faulting, but the most interesting and scenic feature of this part of the rift margin is the line of volcanic hills, craters and cinder cones at Debre Zeit (Bishoftu).

20.3 The Debre Zeit Craters

A satellite image of the craters closest to the town of Debre Zeit is shown in Fig. 20.4. Since the whole line of volcanic features extends over a distance of about 20 km, it is not possible to show them all on an image of this scale. The craters are maars, or explosion craters, similar to El Sod which we saw in southern Ethiopia in Chap. 18, and were formed in a similar manner by a multitude of explosive steam eruptions through groundwater or a shallow lake. The crater rim deposits at Debre Zeit consist of cinders, ash, and fragments of the basalt rock through which the explosive eruptions occurred (PHOTO 20.2). There are about a dozen explosion craters at Debre Zeit, some of which are in double, triple or even quadruple nested groups, as well as numerous cinder cones and hills of rhyolite and pumice, aligned more or less parallel to the Rift Valley. Several of the craters contain attractive lakes, which are becoming increasingly exploited as venues for hotels, picnic gardens and water activities.

The perfectly circular volcano Zuquala also lies on the line of the Debre Zeit cones and craters. Zuquala is formed mainly of trachyte and has a beautiful crater lake at its summit. In a way it is an oddity. It is not part of the line of volcanoes that runs along the centre of the rift but, at about 1 million years old, it is significantly younger than the Pliocene volcanoes Yerer, Wachacha and Furi closer to Addis Ababa. Like the other features noted above, it is connected somehow with this elusive rift margin.

Fig. 20.4 Satellite image showing four of the Debre Zeit crater lakes. The circular craters are maars, formed by violent volcanic explosions. Lake Cheleleka is a swampy, non-volcanic lake. Three other crater lakes: Koftu and Kilole, to the north and northeast respectively, and Aranguadi (*Green Lake*) to the south-southwest, are too far away to show on this scale. Although only four crater lakes are shown on the image, you should be able to make out at least seven actual craters. Image from Google Earth

PHOTO 20.2 Crater rim deposits at Lake Aranguadi, one of a group of maar volcanoes at Debre Zeit. The layers consist of ash and scoria (cindery basalt fragments), and each represents a single explosive eruption. In several places you can see where larger blocks have fallen into the layers, which were probably moist and soggy at the time, and caused them to sag (2006)

20.4 The Rift Margin from Addis Ababa to Lake Chamo

Southwest of Zuquala a distinct rift margin escarpment becomes apparent, its raised shoulder forming the Gurage Mountains (look back to PHOTO 17.1). It is, however, separated from the rift proper by a shallow rifted valley some 2–4 km wide. This type of feature is known as a marginal graben and we will see more of these when we journey along the western margin of Afar later in this chapter.

There are two interesting archaeological sites along the section of the rift margin between Addis Ababa and the Gurage escarpment. About 50 km south of Addis Ababa, in the upper reaches of the Awash River, is the Stone Age site of Melka

Kunture. Here tens of thousands of stone tools (choppers, hammer stones, blades, handaxes, cleavers, flakes), together with remains of *Homo erectus* and the animals which he butchered (mainly hippos), are scattered over an extensive area bordering the river, and a selection of them is displayed in an attractive museum. The stone tool and bone assemblages are interlayered with deposits of volcanic ash, which has enabled the whole sequence to be dated quite precisely. The ages show that these early people lived here almost continually from 1.7 to 0.2 million years ago.

About 30 km south of Melka Kunture, near the village of Tiya, is a field of several dozen standing stone slabs, or stelae, carved from ignimbrite. Many of them are engraved with elaborate designs, including both pagan sun-and-moon, and Christian, symbols. Archaeological excavations have confirmed that this is a burial ground dating from the 12th to the 14th century AD. Although little is known about the people who were buried here, a number of the engravings suggest that they were warriors (PHOTO 20.3).

PHOTO 20.3 Engraved stelae at Tiya. The stones are carved from ignimbrite, and probably mark the graves of warriors. The pair of symbols on the tallest stone is thought to represent two swords, and that below them, like the ones near the base of the two other large stones, a traditional headrest. The chain-like symbol at the broken top of the centre stone may represent the number of people killed by the warrior (2014)

PHOTO 20.4   Ara Shatan maar, "The Devil's Lake", an explosion crater on the western rift margin near Butajira. The crater is about 500 m in diameter at the lake level. The Gurage escarpment can be seen in the far background, and in the foreground is a grey block of basalt, about 1 m across, that has been thrown out in the explosive eruption (2014)

Southwest of Tiya, and behind the town of Butajira, the main escarpment takes on an unusual arcuate shape, almost as if something has taken a bite into the Gurage Mountains. Running along the open edge of the arc is a line of cinder cones, and one solitary and very spectacular maar whose deep crater contains a bright green lake (PHOTO 20.4). It is known as Ara Shatan, or the Devil's Lake. The striking colour of the lake is due to algae which live in its atrophic (oxygen deficient) water.

For a short distance south of the Butajira arc the rift margin is straight, stepped and distinct. Toward Sodo it becomes ill defined once more, and in fact the town of Sodo has a rather similar situation to that of Addis Ababa, in a kind of embayment of the rift. This is probably related, as in the case of the Addis Ababa embayment, to a zone of faults which cuts across the rift margin, known as the Bonga-Goba Lineament. The big Pliocene (or early Pleistocene) volcano Damota, close to Sodo, occupies a rather similar position to that of the Pliocene volcanoes Wachacha, Furi and Yerer, in the Addis Ababa embayment.

Southwest of Sodo a distinct rift margin reappears, but is offset by about 50 km to the east, forming the Chencha Mountains. It continues along the western shores of Lakes Abaya and Chamo before merging into the Broadly Rifted Zone of the south.

20.5   The Southern Margin of Afar

In the region of Sire the eastern margin of the Main Ethiopian Rift begins curving toward the northeast, and finally near Asbe Teferi, between the Chercher and the Ahmar Mountains, it takes a sharp turn to the east to become the southern margin of Afar (Fig. 20.3). As it curves, the step faults of the Sire region break up into many

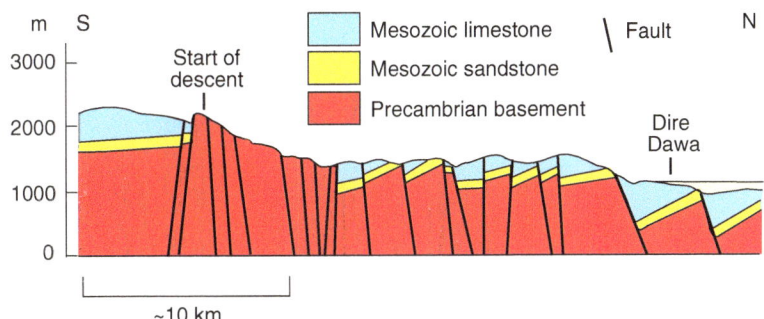

Fig. 20.5 Geological cross section of part of the southern margin of Afar at Dire Dawa, showing the complex faulting. There is not room on the diagram to show the directions of the faults, but it is easy to figure them out from the displacement of the layers. Modified from Geological map of Dire Dawa 1:250,000 (1985), Geological Survey of Ethiopia

smaller faults, and the whole margin becomes broader and more complex. At its eastern end, in the region of Dire Dawa, it consists of dozens of faults (Fig. 20.5). The road which descends to Dire Dawa from the Southeastern Highlands crosses many of these, though most are not obvious to the eye. What is obvious is that one passes the geological formations in what appears to be the wrong order. Precambrian rocks are at the top of the descent (look back to PHOTO 15.1), and for much of the way down, while the much younger Mesozoic limestones are at the bottom. The wide zone of faults has displaced these layers by over 1000 m.

A line of volcanoes (Gumbi, Assebot, Afdem, Gara Adi) runs along the southern margin of Afar. They are similar in age to the volcanoes situated on the eastern margin of the Rift Valley (Kaka, Badda, Chilalo and Encuolo), but unlike them they are located at the foot of the escarpment, not upon its shoulder. How these volcanoes fit into the process of rifting is an intriguing question which as yet awaits an answer.

20.6 The Western Margin of Afar: The Great Escarpment

The highest and most dramatic of the margins is the system of great escarpments, collectively termed the western escarpment, bordering the western side of Afar, shown in Fig. 20.6. It rises between 2000 and 3000 m above the floor of Afar and

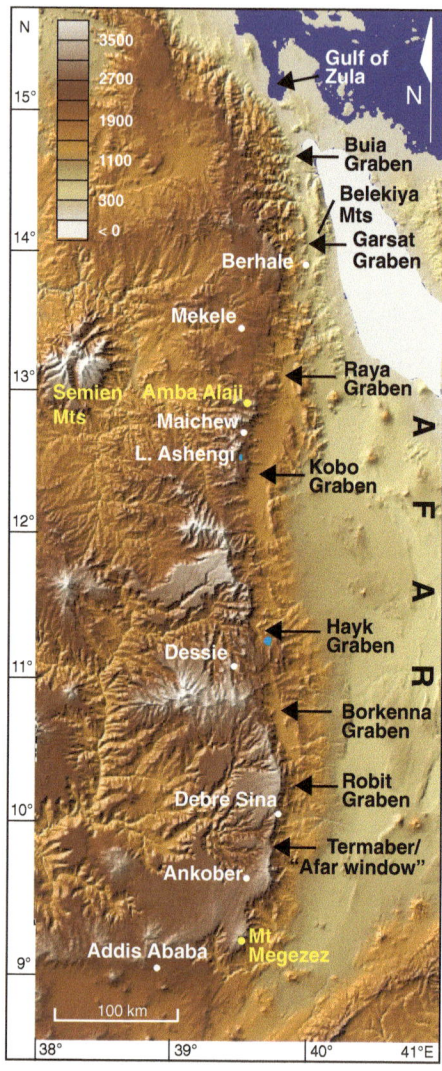

Fig. 20.6 Digital elevation model of the western margin of Afar. DEM from GeoMapApp

in places individual scarps display sheer drops of 500 m and more. Its southern end is marked by Mount Megezez (3457 m), east of Addis Ababa, where it becomes a complex mix of flexures and faults as it curves southwestward to become the margin of the Main Ethiopian Rift. To the north it extends for 600 km, passing through eastern Eritrea and eventually merging with the western margin of the Red Sea. The main road from Addis Ababa to Mekele and northern Ethiopia follows this margin, providing insights into its geology and structure as well as some dramatic scenery.

To begin with, there is a wonderful spot on the high, windy ridge at Termaber, south of Debre Sina, where a stream has cut a gap through the rim of the main escarpment. It is known popularly as the Afar Window. Here, beneath a sheer, vertigo-inducing drop of some 400 m, a broad expanse of eroded foothills descends

PHOTO 20.5 Looking through the "Afar Window" south of Debre Sina. Below a sheer drop of 400 m, rolling foothills extend to Afar which is hidden in the clouds and haze beyond (2011)

to Afar, 50 km beyond and two and a half kilometres below (PHOTO 20.5). Even though Afar itself is generally lost in the distance, mist and haze, the view is terrific!

At Termaber the escarpment drops to Afar in a series of steps and slopes, but a little way north of here its profile changes. Shortly before reaching the town of Debre Sina the road actually cuts through the rim of the escarpment via a 500 m long tunnel, an achievement of Italian and later Ethio-Chinese engineering, then descends through a series of hairpin bends into a kind of miniature rift valley. From here to east of Asmara, in Eritrea, the main western escarpment is bordered by these small rift valleys, each about 10–15 km wide and up to 120 km long (PHOTO 20.6). They can be seen clearly on Fig. 20.6 and their structure is shown in Fig. 20.1(iv). The main escarpment fault faces east, then a series of smaller faults face the opposite way but the faulted blocks are tilted eastward so that the margin still descends, albeit very gradually. Since they occur along the margin, these small rifts have been termed marginal graben. As explained in Chap. 17, a graben is a trough-like valley, similar to a rift valley but on a smaller scale.

PHOTO 20.6   Looking westward across the Robit marginal graben. The ridge just beyond the green area is one of the numerous rhyolite hills that occur within the graben. The main escarpment is hidden in the background haze (2011)

The marginal graben are very obvious to the traveller as their flat floors provide a convenient passage for the main road, and since they are well watered by runoff from the high escarpment they provide fertile land for crops and fruit growing. Sometimes a river flows along them as in the Robit and Borkenna Graben, or they may contain a swamp or a lake, as in the Hayk Graben north of Dessie. Some of the graben have rhyolite hills along their centre line. There are breaks between the individual graben, since they have to follow the offsets of the main escarpment. Where there is an offset, the road climbs over a mountainous ridge to meet the next graben, for example at Dessie between the Borkenna and Hayk Graben. South of Maichew the road turns westward out of the Kobo Graben and ascends the main escarpment to head towards Mekele (PHOTO 20.7). On the way it passes the beautiful Lake Ashengi (PHOTO 20.8) which occupies a small graben higher on

PHOTO 20.7 The road winding over the main western escarpment to Lake Ashengi (2012)

PHOTO 20.8 Lake Ashengi, which occupies a small graben high on the main western escarpment (2012). Photo courtesy Andrew Dakin

the escarpment, and Amba Alaji, an imposing pyramidal massif on the escarpment rim, formed of layers of basalt and ignimbrite (PHOTO 20.9). Amba Alaji provides a natural fortress controlling the pass between Tigray and the Amhara region, and has been the site of several important battles. Notable was that of 1941, which was instrumental in bringing to an end the six-year-long Italian occupation of Ethiopia. Italian troops, believing themselves impregnable in the mountain stronghold, were trapped and defeated by British forces—a reversal of the situation at Amba Aradam six years previously (see Chap. 11).

Along the southern two thirds of the escarpment the graben occur quite high up, about half-way between its foot and its rim. North of Maichew, however, they become offset to the east so that the northernmost ones, the Garsat and Buia Graben, run along its foot. The town of Berhale, on the road to Dallol and the Salt Plain of northern Afar, is situated toward the eastern side of the Garsat Graben. This graben differs in another way from those further south, in that its eastern margin is formed not by an opposing fault but by a tilted block which has been

PHOTO 20.9 Amba Alaji, a natural mountain fortress on the western escarpment. Its pyramid-shaped peak is formed of rhyolite and ignimbrite. Photo courtesy Ernesto Abbate

faulted down from its original level some 2000 m above (Fig. 20.7). This forms a spectacular ridge known as the Belekiya Mountains (PHOTO 20.10).

In the next chapter we will move to Afar, geologically the most intriguing and challenging region of Ethiopia.

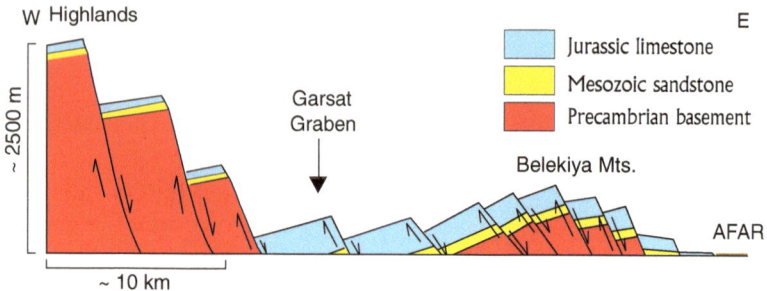

Fig. 20.7   Sketch geological section across the Garsat Graben at approximately the latitude of Berhale. The graben lies between the main escarpment and a tilted, faulted block which has been downfaulted by about 2000 m. Strictly speaking the Garsat Graben is a half-graben, as it is bounded by a fault on one side only. Modified from Explanatory notes to accompany the 1:250,000 Geological Map of Adigrat, by C.B. Garland, Geological Survey of Ethiopia (1978)

PHOTO 20.10   The Belekiya Mountains, taken from the road east of Berhale. The tilted limestone beds are eroded into inverted V-shaped structures known as flat-irons (2011)

The Enigma of Afar

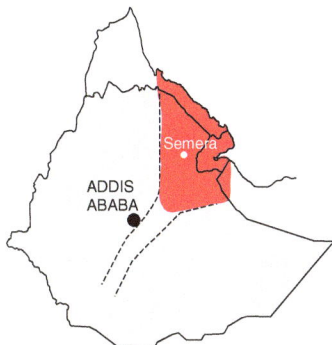

Afar has been variously described as "the lowest place on earth", "the hottest place on earth", "the most hostile place on earth" and even "the hell-hole of creation"! These claims are all somewhat exaggerated, but they do convey an impression of the region—a vast, arid landscape of black volcanic rocks, dusty plains and shimmering salt flats, punctuated by volcanic ranges and massifs and plunging to 153 m below sea level. There is little doubt, however, that to a geologist Afar must be one of the most fascinating places on earth, as it is the only place where the final stages of the break-up of a continent and the early stages of sea-floor spreading can be seen on land—actually happening!

So what exactly is Afar? The terminology regarding it is probably more confusing than any other in Ethiopia. Ethnic regions, political domains and international boundaries have become confused with geographical and geological ones, resulting in a hotchpotch of terms that are generally used without explanation. Here are just a few examples: Afar Triangle, Afar Rift, Afar Depression, Danakil, Danakil Depression, Afar Desert, Danakil Desert, Dankalia... and so forth.

Since these terms are not clearly defined, I will avoid most of them. The term "Afar" will be used in the following sections to refer to the whole region bounded by the foot of the western escarpment, the foot of the southeastern escarpment and the western coast of the Red Sea as shown in Fig. 21.1 and the thumbnail map at the

© Springer International Publishing Switzerland 2016   243
F.M. Williams, *Understanding Ethiopia*, GeoGuide,
DOI 10.1007/978-3-319-02180-5_21

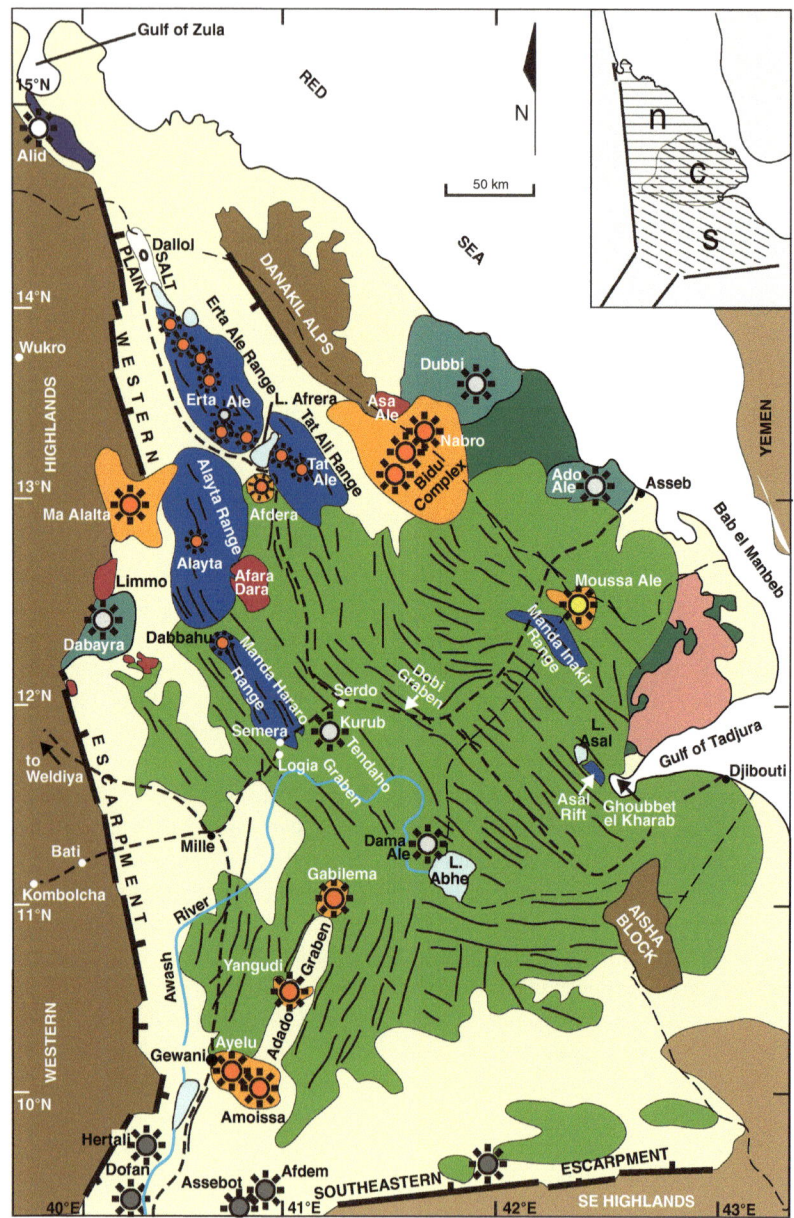

Fig. 21.1 (continued)

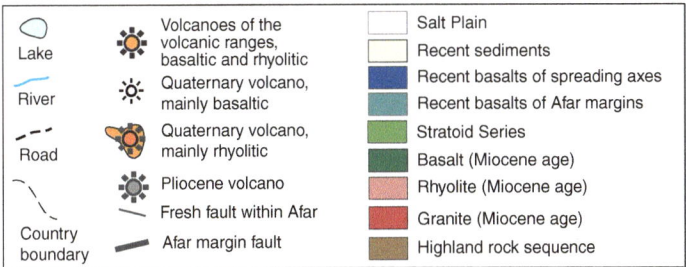

◀ **Fig. 21.1** Geological map of Afar. The *inset* shows the informal northern (n), southern (s) and central (c) regions explained in the text. Faults are shown schematically as there are too many to show in detail. Based on 1:500,000 maps of northern Afar (prepared by J. Varet et al.) and central and southern Afar (prepared by the CNR-CNRS Afar team). Produced by CNR-CNRS 1973 and 1975 respectively

top of this chapter. This includes the whole of Djibouti and the coastal strip of Eritrea, which are geographically, geologically and even ethnically a part of Afar and cannot be excluded from this and the following chapters. Because the Red Sea spreading axis, the Gulf of Aden spreading axis and the African Rift System all terminate within Afar, it is sometimes referred to as the Afar Triple Junction but this term should, strictly speaking, apply to the more localised region to which the three axes converge. In Chap. 23 we shall look into where the actual triple junction is.

Afar is roughly the shape of a right-angled triangle whose sides are defined by the southeastern and western escarpments and hypotenuse by the Red Sea coast. It measures, in very round figures, 350 km from east to west along its southern margin and 600 km from south to north, and slopes generally from south to north—from about 1000 m above sea level near the foot of the southeastern escarpment to just over 120 m below sea level in the north. It reaches its lowest point near the tip of the Gulf of Tadjura in Djibouti where the salt lake Asal, at 153 m below sea level, is the lowest place not only in Afar but in the whole of Africa, and the second lowest place in the world on land (after the shores of the Dead Sea). Although summer temperatures in Afar can reach over 50 °C (according to one report up to 67 °C!), winter temperatures closer to 30 °C are perfectly manageable for a visit to this fascinating part of Ethiopia.

Before the days of air travel it was necessary to cross this inhospitable, and in those days highly dangerous, region by foot or camel in order to reach the Western Highlands from the Red Sea coast. Afar thus presented a formidable obstacle for would-be visitors to what was then the Ethiopian (or Abyssinian) heartland, and the few brave souls who attempted to explore the region itself frequently met with disaster due to thirst, heat, or massacre by the warlike and xenophobic inhabitants. For example, the Italian explorer Guiseppe Giulietti's entire expedition of 33 people was killed by Afar tribesmen in 1881. Despite the discouraging conditions, however, a surprising amount was achieved. The few intrepid travellers who did trickle through made a great many useful geological observations as well as some preliminary maps, and the Italians, during their colonisation of Eritrea and later brief occupation of Ethiopia, made some outstanding contributions.

What really put Afar on the geological map, however, was the development and gradual acceptance of the plate tectonic theory during the 1960s. Geologists quickly realised that Afar played a crucial role in the plate tectonic scheme of things. French, Italian, German and other teams descended on the region and carried out detailed mapping and some geophysical studies, though still under very difficult conditions. More recently, the development of sophisticated research tools and vastly improved accessibility to the region have led to a kind of research onslaught on Afar, and both Ethiopian and international teams of earth scientists have been using every means possible to find out what is going on there.

Although Afar is a geological and geographical entity it is convenient to sub-divide it into northern, central and southern regions, firstly because a single chapter covering the whole of Afar would be much too long, and secondly because the regions do have different geological characteristics. For convenience I will take the boundary as being very roughly the line of the highway which traverses Afar from Bati on the western escarpment to Asseb on the Red Sea Coast[1] (Fig. 21.1). Southern Afar is considered to be more or less to the south of the highway, northern Afar to the north of it, and a broad region where they overlap is regarded as central Afar. This nomenclature, illustrated in the inset to Fig. 21.1, is entirely informal, as signified by the use of lower-case letters when referring to "northern", "central" and "southern".

Figure 21.1 shows a simplified geological map of Afar. It might be a good idea for you to bookmark this page as it will be referred to a number of times in this and the following two chapters. You can see right away from the colours on the map that the geology of northern Afar differs from that of the southern and central regions. Northern Afar is characterised by volcanic ranges which are shown in blue. These ranges consist of fresh lava flows topped by very young volcanoes, several of which are active, and dissected by fissures and faults, all aligned roughly NNW to SSE Southern and central Afar are characterised by flat-lying layers of slightly older volcanic rocks, mainly fresh-looking black basalt, shown in green on

---

[1] At the time of writing (2015) this highway is closed to through traffic at the Eritrean border.

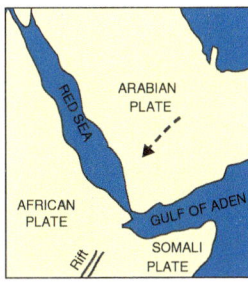

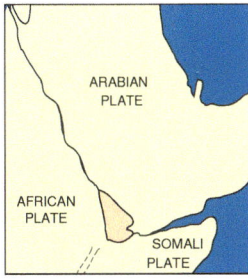

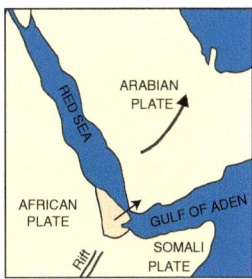

(i) Imagine the Arabian Plate to be moved back to where it was before the Red Sea opened.

(ii) Everything would fit back neatly, except the Gulf of Aden, which had already begun to open, and Afar, shown in darker yellow, where there would be an overlap.

(iii) Afar is still partly attached to the Arabian Plate as the Red Sea opens, and is being pulled and stretched as the plates try to make their final break.

Fig. 21.2 How Afar may have formed when the Arabian Plate got stuck!

the map. These layers are cut by hundreds of faults forming valleys and ridges, horsts and graben. In the valleys and graben, and surrounding the volcanic ranges of northern Afar, are flat plains covered by sand and salt—the products of long dried up lakes, rivers and, in northern Afar, the sea. The brown areas on the map represent the highlands, which in places rise to over 3000 m above the Afar floor.

Afar is an enigma—it really shouldn't be there or, since it obviously is, it should be ocean. If you imagine Arabia and Africa to be fitted back together as they were before they separated, the southern corner of the Arabian Peninsula (now Yemen) would slot neatly into the corner where Afar is now (Fig. 21.2). As Arabia moved away, it looks as though the Red Sea should have continued to open southwards, creating new ocean floor to connect with that of the Gulf of Aden. Instead of this a piece of Africa—Afar—got stuck to the corner of Arabia and was pulled out and stretched as it moved away, and is only now reaching breaking point.

In the following two chapters we will look at the geological features of northern, southern and central Afar, and see how they enable us to gain an insight into how and why continents break apart and how ocean floor begins to form between them.

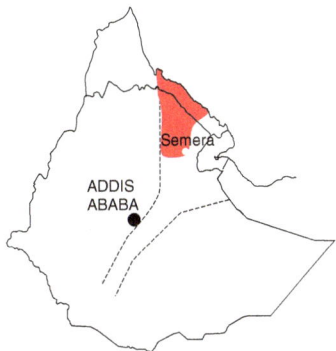

So much is happening in northern Afar that it will be helpful to begin with a brief recapitulation.

Figure 22.1 summarises the events that are going on in and around Afar today. The Arabian Plate is moving away from Africa in an anticlockwise rotation with a hinge at the northern end of the Red Sea, and is colliding with the Eurasian Plate to form the Zagros Mountain belt. At the same time both Africa and Arabia are moving slowly northwards. Sea-floor spreading is taking place along the Gulf of Aden, and its spreading axis has begun to penetrate the African continent along the Gulf of Tadjura in Djibouti. The Red Sea is also widening. A spreading axis has developed in its southern section, and has been gradually growing to the north and south. The Somali Plate has begun to pull away from the African Plate, but has not really separated from it as no ocean floor has yet been created between them. The Main Ethiopia Rift is widening at a much slower rate than the Red Sea and Gulf of Aden and is still entirely underlain by continental crust.

Toward the southern end of the Red Sea, at a latitude of about 14°N, its spreading axis appears to terminate. Although a narrow seaway connects the Red Sea and the Gulf of Aden along the strait of Bab el Mandeb, this is underlain by continental crust and the extreme eastern part of Afar is actually an extension of the Arabian Plate! But in fact the Red Sea spreading axis has not terminated—it has

© Springer International Publishing Switzerland 2016
F.M. Williams, *Understanding Ethiopia*, GeoGuide,
DOI 10.1007/978-3-319-02180-5_22

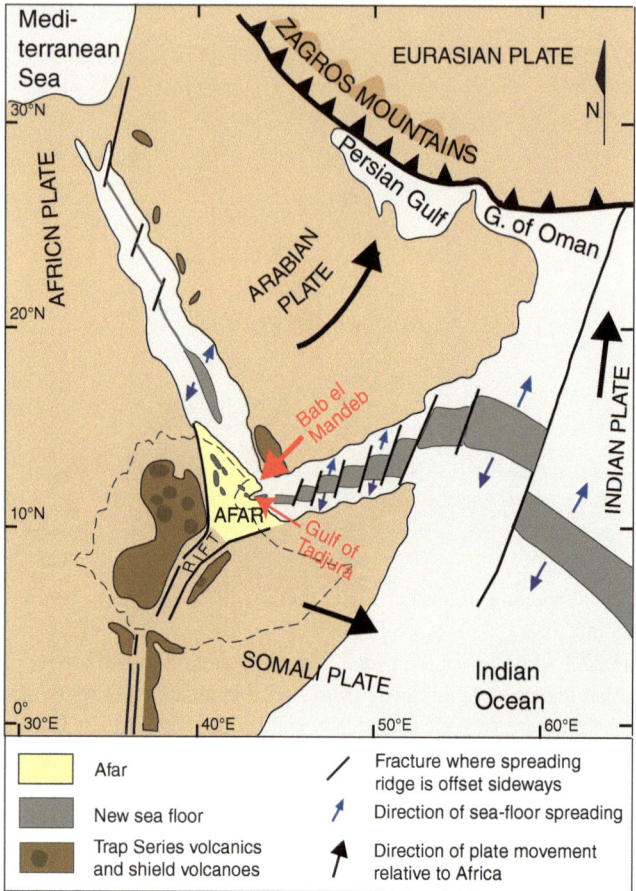

Fig. 22.1 The current situation. The *large black arrows* represent the direction of movement of the plates, the *small blue arrows* the direction in which the spreading axes are opening. The Red Sea and the Gulf of Aden are widening at about 2 cm per year and the northern part of the Main Ethiopian Rift at 0.5 cm per year. Note that everything south of the subduction zone shown by the *toothed line* is moving northwards toward the Eurasian Plate. This general northward movement is not shown on the figure; the *black arrows* show movements relative to the African Plate (as though the African Plate were standing still). Notice also that the spreading axes are not opening at right angles to the general direction of the axes. The axes consist of segments which are offset from each other in en échelon fashion, and are connected by fracture zones marked as *thin black lines*

simply jumped sideways into Afar and split into a series of mini spreading axes. These are the volcanic ranges[1] that are shown in dark blue on Fig. 21.1.

The composition of the basalts erupted along these volcanic ranges confirms that something unusual is going on here. Oceanic basalts have subtle but significant differences in composition from those produced on continents, such as the Trap Series. The basalts of the volcanic ranges are not exactly like either, but are something in between. They are an early sign that a new ocean floor is being born —on land!

22.1 The Volcanic Ranges

There are four volcanic ranges in northern Afar: Erta Ale, Tat Ale, Alayta and Manda Hararo (see Fig. 21.1). They are offset from each other, and are oriented in a NNW to SSE direction, parallel to the Red Sea. Each consists basically of a ridge of basalt which erupted through fissures over the past million years or so, topped by one or more central volcanoes and dissected by many later fissures and faults. The most recent fissure lavas erupted during the past few thousand years, right up to the present time. The volcanoes are either active or dormant, showing signs of current or recent activity in the form of fumaroles, hot mud pools and very fresh lava flows. Erta Ale volcano has two active lakes of molten lava, and in the Manda Hararo range there was a recent major volcanic eruption and formation of new fissures.

22.2 The Erta Ale Range

A geological map of the Erta Ale range is shown in Fig. 22.2. It is built upon a long, low ridge of basalt that has erupted through fissures now hidden beneath it. Although the oldest lavas may be up to a million years old, most are much younger and the most recent ones are only a few years old. The early fissure eruptions that built the ridge changed gradually to eruptions through vents, forming central

---

[1]These are sometimes referred to in the geological literature as axial ranges, volcanic segments or magmatic segments, but for consistency I will stay with the descriptive term volcanic ranges.

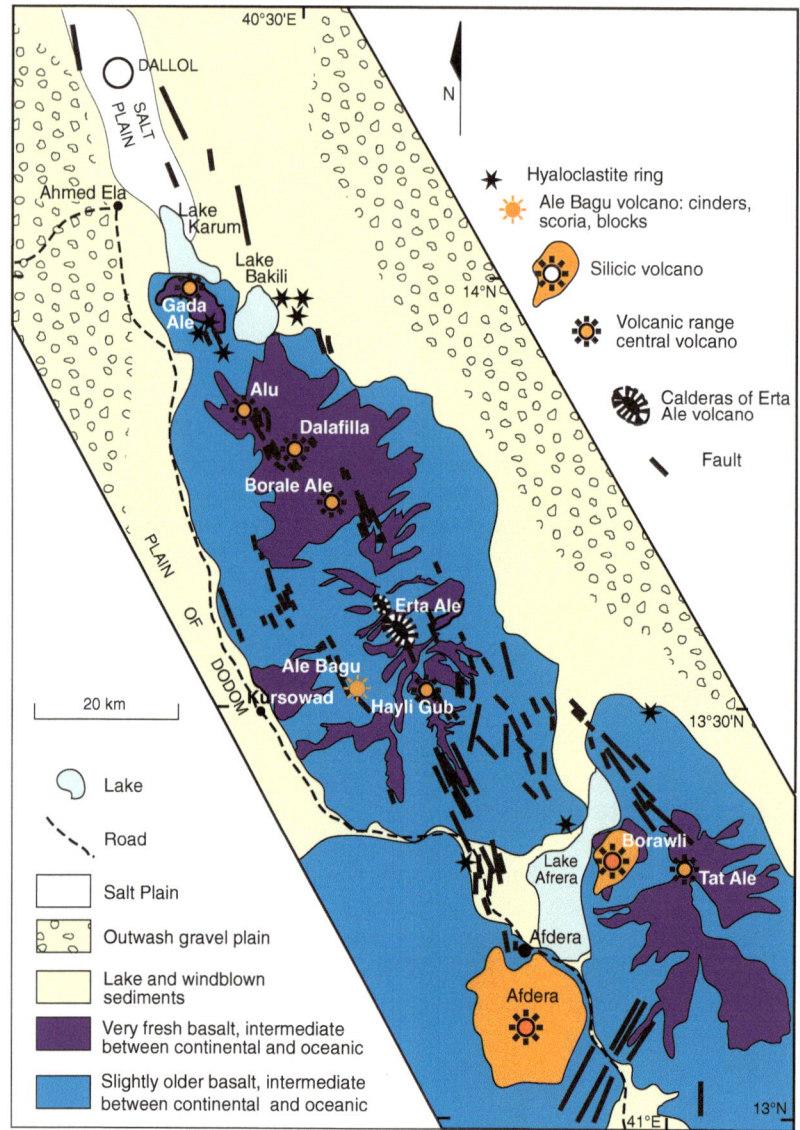

Fig. 22.2 Geological map of the Erta Ale Range. Based on the 1:500,000 Geological map of northern Afar (prepared by J. Varet et al.) and 1:100,000 Geological map of the Erta Ale Volcanic Range (prepared by F. Barberi and J. Varet). Produced by CNR-CNRS 1973 and 1972 respectively

PHOTO 22.1 Track to Erta Ale, the low ridge in the distance. The flows of basalt have erupted from the fissure system on which the volcano is located. To reach the volcano one is walking over the very early stage of a new ocean floor—forming on land! (2013)

volcanoes superimposed on the ridge. These are, from north to south: Gada Ale, Alu, Dalafilla, Borale Ale, Erta Ale, Ale Bagu and Hayli Gub. All except Ale Bagu lie on the same line of fissures. Ale Bagu lies on a separate line, displaced slightly to the west, and it differs from the other volcanoes in that some of its products are not lava flows but deposits from explosive eruptions: scoria, cinders and blocks of light and dark-coloured rock. The other central volcanoes are built of lava flows. Their earlier flows were of basalt but, with the exception of Erta Ale, their later eruptions produced rocks richer in silica, often a trachyte or rhyolite pitchstone which is easily mistaken for basalt on account of its dark colour. All the volcanoes show abundant fumarolic activity, Gada Ale has a pool of boiling mud and Erta Ale itself its famous lava lakes. Faulting and fissuring have taken place throughout the building of the range, and fresh faults cut even the latest lava flows. In 2008 a new fissure opened east of Alu and Dalafilla volcanoes, with the eruption of a flow of basalt lava.

22.3   Erta Ale Volcano

This is Ethiopia's most famous volcano, and a rarity in the whole world for it is one of only four[2] volcanoes that have an active lake of lava in their summit crater. Seen from a distance, however, it is not an imposing mountain (PHOTO 22.1). Although

---

[2]This number varies between three and seven depending upon the source one consults.

PHOTO 22.2 Braids of ropey lava along the track to Erta Ale (2013)

it is the highest volcano in the range, its summit rises only 675 m above the level of
the surrounding plain, which is itself 60 m below sea level. However, the contorted
flows of fresh black lava which one passes on the fairly gentle (but hot!) walk to
the summit (PHOTO 22.2) are spectacular indeed and give a taste of what is to
come.

Figure 22.3 shows an aerial photograph of Erta Ale. At the summit are two
large, elliptical calderas which are filled with very fresh basalt (PHOTO 22.3). The
larger caldera, to the south, is about 3 km long and the northern one about 1.5 km.
The northern caldera contains two circular pit craters, each of which contains a lake
of molten lava.

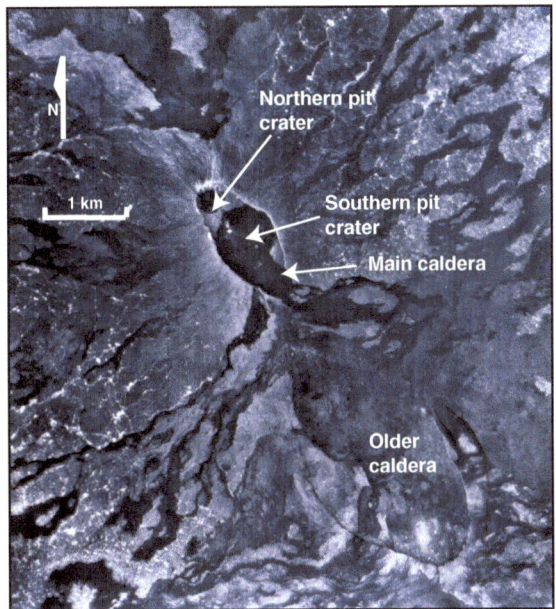

Fig. 22.3 Aerial photo of Erta Ale volcano

PHOTO 22.3 Erta Ale's smaller summit caldera, floored by very fresh basalt probably produced by overflow from the lava lakes as well as from fissures (2013)

PHOTO 22.4 Erta Ale's southern pit crater and lava lake. The temperature of the molten lava is about 1080 °C. The pit is about 65 m in diameter (2013)

The activity of the two lava lakes is constantly changing, and the description given here is only a snapshot of that seen most recently by the author, in February 2013. At that time the southern pit crater was the more active. A solid crust had developed over the lava in the northern crater and a 10 m high hornito, or pillar of solidified lava, had built up near its centre. This occasionally shot out a stream of red lava, indicating the presence of molten material not far beneath. The lava in the southern pit crater, however, was in perpetual motion beneath a thin surface crust which continually cracked and split open to reveal the red molten material beneath (PHOTO 22.4). The slabs of crust moved across the surface of the lava lake, pushed apart by rising lava and sliding beneath each other, just like plate tectonics in miniature! Every so often the lava burst through the crust as a fiery fountain (PHOTO 22.5). This appears to be the most widely reported scenario over the past two decades. Reports from the early 1970s, however, indicate that the two lakes were equally active, and one from 2004 reports that both had formed a solid crust and the only activity was in the form of glowing hornitos. Like all volcanoes, Erta Ale is changeable and unpredictable!

The levels of both lava lakes vary considerably, and not synchronously, from about 200 m below their crater rims to the very top, and occasionally they overflow into the caldera. The solidified lava around the crater rims has a greenish, rather furry looking coating. This is formed of thin drawn-out strands of volcanic glass thrown out by lava fountains and is called Pele's Hair, after the Hawaiian goddess of volcanoes.

It is not known for certain how long Erta Ale's lava lakes have been active. The German explorer Johann Maria Hildebrandt made the earliest written report of an

PHOTO 22.5 Lava fountain at night, at the edge of Erta Ale's lava lake (2013)

ascent of Erta Ale, in 1873, but his description suggests that he ventured no further than the rim of the main caldera. It is often quoted in the literature that the first European to observe one of the lava lakes was the Italian explorer and adventurer Tullio Pastori, in 1906.[3] However, he left no written record of his visit and his word-of-mouth report, made more than 60 years later to geologists of the Italian-French mapping mission, was regarded by some with scepticism. The first reliable report of an active lava lake was by the French and Italian geologists themselves, in 1967. Since then the lava lakes have been observed on a regular basis and, with fluctuations as described above, their activity has been continuous.

Another intriguing question is what keeps the lava molten and active, continually rising and falling within its vents, without solidifying completely as lava

---

[3]There are fascinating stories associated with both these explorers. Hildebrandt, for example, having suffered extreme hardships on his ascent and descent of the volcano, and with his feet swollen from the heat, war furious to return to his camp and find that his companions had eaten all the food. Pastori claims to have crossed Afar from west to east taking with him only water and an onion to quench his thirst!

normally does when it reaches the earth's surface. I have been unable to find a satisfactory answer to this question, but it probably relates to there being a rare and delicate balance between the depth of the magma chamber from which the molten lava comes, the pressure within the earth at that depth, the pressure at the base of the lake itself, and possibly the amount of gaseous (bubbly) material in the lava.

22.4 An Ancient Sea Bed

The plains surrounding the Erta Ale Range, at least as far south as Lake Afrera,[4] are between 50 and 100 m below sea level. To the north they descend even lower, reaching 130 m below sea level west of Dallol. As we will see in the next section, the sea has overflowed from the north into this region on several occasions in the past, and probably completely surrounded the Erta Ale Range. There is evidence for this in the form of features known as hyaloclastite rings. If you look back to Fig. 22.2 you will see these marked as black stars on the map. They are small, rimmed craters formed as a result of explosive eruptions, rather like maars except that their rim deposits contain a great many glassy fragments (*hyaloclastite* means, literally, "glassy broken pieces"). The hyaloclastite rings in northern Afar have shallower craters than maars (whose craters are generally deeper than the surrounding land), and their rims are very level so that they appear to be flat-topped when viewed from the side. They closely resemble features known as guyots, or sea-mounts, which are typically found beneath oceans, and whose flat tops are believed to be the result of decapitation by wave action. Although a physical resemblance does not prove that they are guyots, the presence of fossil marine corals on their flanks and summits is fairly clinching evidence that they formed beneath an ancient sea which, north of the Erta Ale Range, has left its indisputable mark in the form of a great salt-covered plain.

22.5 The Salt Plain

The Salt Plain is a vast expanse of white, shimmering salt (PHOTO 22.6) which extends some 40 km northwards from the tip of the Erta Ale Range, and mostly lies at around 120 m below sea level. It is the product of numerous past invasions of the sea, which flowed in from the north via the Gulf of Zula and subsequently

---

[4]Lake Afrera used to be called Lake Giulietti after the unfortunate explorer mentioned in the previous chapter.

PHOTO 22.6 The Salt Plain of northern Afar (2013)

evaporated to leave behind thick layers of salt. These layers consist not only of common salt (sodium chloride) but also salts of potassium, mainly sylvite (potassium chloride), which are valuable for the manufacture of fertilisers. Mining of potassium salts, collectively known as potash, from the Salt Plain has been carried out since the mid-1920s by Italian, Ethiopian and American companies, but intermittently and on a relatively small scale. Recently, however, potash mining has recommenced on a large scale, raising concerns about its possible effects on this fragile and hitherto remote environment.

Some of the early mining operations and related geological surveys have provided useful information about the thickness and nature of the salt deposits. Exploratory drillholes of the Ralph M. Parsons Company of California, which operated in the area during the 1960s, penetrated 975 m of salt, and geophysical evidence (seismic and gravity measurements) has indicated that it could be a great deal thicker, possibly as much as 3 km! This great thickness was made possible because the floor of the Salt Plain was subsiding, much as the floor of the Red Sea had subsided to form a salt-filled trough some 25 million years earlier as Arabia began moving away from Africa (see Chap. 8).

Since the Salt Plain covers an area of some 400 km^2 and has a thickness of 975 m at the very least, there is a lot of salt there! It has long been extremely valuable to the local economy—and it is a renewable resource. Every year, during the rainy season in the highlands, mud-laden rivers pour down the western

PHOTO 22.7 Layers of salt and mud that overly the main salt deposit (coin for scale) (2012)

escarpment, covering the Salt Plain to depths of up to a metre and raising the level of the groundwater table. Dissolved salt rises to the surface and is precipitated when the water evaporates, forming a surface crust of interlayered salt and mud which is renewed every rainy season (PHOTO 22.7). Over the years the layers of salt and mud have reached a thickness of around 4 m. The true salt beds, originally deposited when the sea flooded the region, are completely concealed beneath them. Mining of the surface salt layers has been an important local industry for centuries (PHOTO 22.8), and salt blocks or "amole" used to be a form of currency and are still a valuable trading commodity. The salty crust is expertly cut into blocks of equal size (PHOTO 22.9) and loaded onto camels and donkeys which carry it up to market in the highlands—a journey which takes about five days (PHOTO 22.10).

The two lakes, Karum[5] and Bakili, at the southern end of the Salt Plain where it abuts against the Erta Ale Range, are remnants of the floods from the highlands.

---

[5]Lake Karum is sometimes also known as Lake Asal, resulting in confusion with the lake of that name in Djibouti.

PHOTO 22.8 Salt-mining on the Salt Plain of northern Afar (2013)

PHOTO 22.9 Salt cutter skillfully fashioning blocks ("amole") from the crust of interlayered salt and mud overlying the main salt deposit, northern Afar (2013)

PHOTO 22.10 Camel train carrying salt, with the foothills of the western escarpment in the background. In the *foreground* is an outwash gravel plain, covered with stones dumped by rivers flowing from the escarpment. Due to the recent construction of a tarmac road ascending the escarpment, these picturesque camel trains may become a thing of the past (2011)

Because they are located in a dip, their water does not evaporate completely but it is very salty. Lake Karum is the saltiest lake in northern Afar.

Fossil marine corals are found on the hillslopes bordering the Salt Plain, and study of these indicates that the sea has flooded the plain at least three times, about 200,000, 120,000, and 80,000 years ago. The Salt Plain would be beneath the sea now, had a lava barrier near Alid volcano, south of the Gulf of Zula, not blocked its inlet. It has been dry, apart from the annual floods, for the past 35,000 years.

22.6   Dallol

Within the Salt Plain are regions of intense geothermal activity in the form of hot springs, intermittent geysers, brine pools in which carbon dioxide bubbles to the surface, bubbling mud pools, and circular crater-like structures which are collapsed mounds of salt blown up by steam eruptions. This activity is really an extension of that of the Erta Ale Range, and in fact geophysical studies have indicated that the range itself continues northward as a volcanic ridge buried beneath the salt.

The best known of these geothermal manifestations is Dallol. This is essentially a volcano buried beneath several hundred metres of salt which it has pushed up into an elongated dome, some 5 km long and rising to about 50 m above the surrounding plain. From time to time the volcano degasses itself, and the top of the

PHOTO 22.11 Salt mound in Dallol caldera (2012)

dome has collapsed to form a broad, shallow depression in which springs of heated, salt-saturated groundwater and magmatic fluids bubble up to form a colourful landscape of salt mounds, fumaroles and warm pools (PHOTO 22.11). When a new spring forms it precipitates white salt, but as the mound grows its outer parts dry out and the small amounts of iron mixed with the salt become oxidised. The bright colours are due to various forms of iron oxide and iron hydroxide mixed with the salt, as well as some deposits of sulphur. Individual springs eventually become blocked with salt, rendering them inactive, and new ones form to replace them. The landscape is continually changing as springs die and new ones grow elsewhere, and pools of red, orange or bright green acid water dry out, then re-appear (PHOTOS 22.12 and 22.13). In some spots fluctuating water levels have resulted in miniature salt formations which resemble flowers and mushrooms (PHOTO 22.14).[6] It has been suggested that Dallol is equivalent to the

---

[6]At least, this is way it used to be. Over the past few years the formations at Dallol have deteriorated markedly. The reasons for this are controversial.

PHOTO 22.12 Green lake at Dallol (2012)

geothermal vents known as black smokers that are found beneath fully developed oceanic spreading axes such as the Mid-Atlantic Ridge.

In the northern part of the Dallol dome the salt layers have been eroded into spectacular formations of towers and cliffs, resembling the castles of fairy tales. The pinnacles are capped by a relatively hard layer of gypsum, anhydrite (calcium sulphate) and clay, which protects the material beneath from erosion (PHOTO 22.15).

22.7   The Other Volcanic Ranges

The Erta Ale range represents the northernmost segment of the offset Red Sea spreading axis. At the southern end of this range the spreading axis splits into two: the Tat Ale and Alayta ranges. At the splitting point, nestled between the three ranges, is the salt lake Afrera which is fed by hotsprings and by rare run-off from

PHOTO 22.13 The same locality as PHOTO 22.12, taken a year later from a slightly different angle. The *green lake* has disappeared and a field of *red iron-stained salt crystals* has taken its place (2013)

the surrounding mountains, and a big volcano called Afdera. Afdera does not belong to any of the volcanic ranges. It is older than they are, and is built of lavas of varying composition, ranging from basalt to silica-rich rhyolite. The Tat Ale and Alayta ranges are similar in structure to the Erta Ale range, being composed of fresh basalt lavas erupted through fissures and vents, and later eruptions of more silica-rich types. There are no reports of recent eruptions from Tat Ali, but Alayta volcano erupted in 1907 with the extrusion of a large lava flow, and again in 1915.

22.8 The Manda Hararo Range: Rifting in Action

The Manda Hararo Range is particularly interesting because tectonic and volcanic activity there have occurred recently enough to be monitored using state-of-the-art geological, geochemical and geophysical techniques. Like the ranges to the north,

PHOTO 22.14 Salt "mushroom" at Dallol, caused by variations in the water level. *Pen* for scale (2011)

PHOTO 22.15 Salt castles near the northern end of Dallol mountain (2011)

PHOTO 22.16 The fissure-vent, 60 m deep, that opened on the eastern flank of Dabbahu volcano in September 2005. The photo was taken looking toward the north, and the *hill* in the *background* is a small volcanic cone called Da'Ure, located adjacent to Dabbahu (2008). Photo courtesy Lorraine Field

it is a broad ridge formed of fissure basalts, very much dissected by faults and open fissures oriented approximately NNW to SSE. At its northern end is a complex volcano known as Dabbahu (also referred to as Boina) formed of dark glassy trachyte, pumice and abundant flows of obsidian, perhaps all overlying a now concealed basalt shield volcano.

In September 2005 an unusual amount of earthquake activity was recorded at the seismic station in Addis Ababa, and study of the records showed this activity to be concentrated in the region of Dabbahu volcano. Three weeks later a vent, and a fissure 500 m long and 60 m deep, suddenly opened on the eastern flank of the volcano (PHOTO 22.16). Local camel herders who witnessed the event described it graphically as "a violent explosion, black rocks that flew like birds, an umbrella-shaped cloud and ash that darkened the sky for three days". Examination of radar data that had routinely been collected by satellites prior to and following the event showed that a strip of land about 60 km long, running southwards from Dabbahu and along the path of the new fissure, had bulged upwards by about 1.5 m since the earthquake activity began. When the vent and fissure formed, it subsided along its central axis by about 2 m. Clearly something very interesting was going on, and a major project was quickly organised to put in place a network of seismometers and GPS stations to monitor events, and to carry out gravity surveys and field studies.

Modelling of the data obtained from these studies showed that the vent and fissure were the surface manifestations of a 60-km-long basalt dyke, 8 m wide and 10 km deep, which had ruptured practically the entire length of the Manda Hararo Range from Dabbahu volcano almost to the main Bati-Asseb highway. Over the following five years a further 13 dykes, much shorter than the first one, were emplaced, and in 2007 and 2009 eruptions of basalt occurred south of the first dyke. As well as more than 8 m of extension produced by dyke injections themselves, the crust in between them stretched by about 1 m across the range. The Manda Hararo Range is now 9 m wider than it was before! At the time of writing the activity appears to have stopped but, if it does resume, the fissure and the accompanying volcanic and earthquake activity are heading straight for Semera, the administrative capital of the Ethiopian Afar Region.

22.9   The Danakil Alps

Between the Erta Ale range and the Red Sea is what appears at first sight to be a misfit: a ridge of Precambrian basement rocks overlain by Mesozoic sedimentary rocks, similar to the sequence found on the highlands to the west (see Fig. 21.1). It is a substantial feature, about 150 km long, 50 km wide and 400 m high and has been called, rather imaginatively, the Danakil Alps. It is sometimes also referred to as the Danakil Horst, but it is not a true horst as it is bounded by a fault only on its western side. Rather than being an uplifted block it is a piece of the Western Highlands which was stretched out as the Arabian Peninsula moved eastwards, and finally broke off as the Red Sea spreading axis jumped sideways to form the Erta Ale range. Because it is separated from the African Plate by the Erta Ale range, and from the Arabian Plate by the southern end of the Red Sea spreading axis, it really belongs to neither. A number of researchers regard it as a little plate in its own right, and refer to it as the Danakil Microplate.

22.10   Other Volcanoes in Northern Afar

It is clear from the map in Fig. 21.1 that not all the volcanoes in northern Afar are part of the volcanic ranges, or are composed mainly of basalt. Ma Alalta, Afdera and the three volcanoes of the Bidu Complex which straddles the border with Eritrea, are all formed mainly of silica-rich rhyolite. They are big volcanoes—and are explosive, unlike those of the volcanic ranges. In 2011, for example, Nabro, a member of the Bidu complex located just over the Eritrean border, erupted violently, killing 31 people and emitting lava flows and an ash cloud that spread over hundreds of kilometres and disrupted air traffic for several days. These volcanoes appear to be misfits in this mainly basaltic terrain, until you notice that, with the exception of Afdera, they are located near marginal regions of Afar which are still underlain by continental crust. Study of the chemistry of their lavas confirms that their composition has indeed been influenced by their passage through silica-rich continental rocks. It is possible that E-W faults intersecting the margins of Afar have facilitated volcanic eruptions at these particular locations.

Unfortunately this explanation cannot account for Afdera, which sits pretty much in the centre of northern Afar. Its presence is as yet unexplained. Possibly it belongs to a rift which ceased to develop when the Erta Ale range split into the Tat Ale and Alayta ranges. It is one of the many Afar puzzles awaiting solution.

22.11   How Old Is Afar? The Red Series and Granite Intrusions

In a few places at the western edge of Afar, where the foothills of the western escarpment begin rising from the flat plain, are layers of bright red sand and gravel known as the Red Series (PHOTO 22.17). They are thought to be the very first sediments deposited in Afar, washed down from the newly-forming escarpment. Fortunately they are interlayered with some flows of basalt which can be dated radiometrically. Those at the base are dated at 24 million years and those at the top at 4 million years.

If you look back to Fig. 21.1 you will notice some small red blobs, three of which are labelled Limmo, Afara Dara and Asa Ale. These are granite intrusions dated at 25 million years old. It is thought that they formed from magma intruded along the early faults of the western escarpment and were exposed as faulting progressed. They became separated as Afar stretched and widened and the volcanic ranges formed between them.

PHOTO 22.17 The Red Series—*red sands* at the foot of the western escarpment, exposed in a dry stream bed. These are thought to be the earliest sediments deposited in Afar (2011)

Both these observations suggest that the faulting along the western escarpment which led to the formation of Afar commenced about 25 million years ago. That was also the time that the Red Sea was forming as an elongated trough (see Chap. 8), thus supporting the idea that Afar is a part of the Red Sea spreading zone.

In the next chapter we will move to southern and central Afar, which hold not only further clues about how a continent breaks apart but also the key to where we humans came from.

Southern and Central Afar: Lava Flows and the Birthplace of Mankind

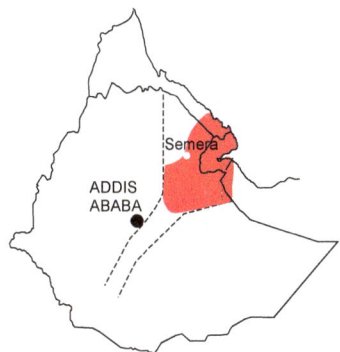

Southern and central Afar offer a volcanic landscape unlike anything else in Ethiopia, or perhaps in the world, and present the geologist with some intriguing and challenging questions. Whereas in northern Afar it is possible to see much of what is happening, right at the earth's surface, the events that led to the landscapes, rock formations and complex structure of the region further south are, though immensely significant, less easy to figure out.

The region's greatest claim to popular fame is its role in revealing the early history of mankind, for here the remains of our early ancestors have been found in abundance, leading to its reputation as "the birthplace of mankind". We will visit this topic later in this chapter, but first we will turn our attention to the geology, beginning with its most prominent and in my opinion most puzzling feature:—the volcanic rocks which cover most of southern and central Afar known as the Stratoid Series.

23.1 The Stratoid Series

The rocks of the Stratoid Series are shown in green on the map of Fig. 21.1. They consist of flat-lying layers mainly of basalt, with some ignimbrite and rhyolite, and occasional interbedded sediments showing that there was sometimes sufficient time between eruption of the lava flows for lakes to form. These volcanic rocks were presumably erupted through fissures which are now buried beneath the flows. In some ways they resemble the Trap Series of the highlands but they are much younger, between 4 million and half a million years old. Their composition is also different, and resembles that of the basalts of the volcanic ranges—not quite continental and not quite oceanic. Many of the basalt flows show spectacular columnar jointing (PHOTO 23.1). The tops of the columns weather into cannonball-like boulders which frequently topple onto the highway, creating an additional traffic hazard to those posed by the hairpin bends circumventing the fault scarps, and the soporific desert heat.

The layers of the Stratoid Series are cut by hundreds of faults, resulting in a landscape, sometimes described as lunar, of black rocky ridges and boulder-strewn valleys, sometimes with a veneer of sandy sediment blown in by the wind. Patches of thorny scrub and tough hummocky grass provide sparse fodder for the goats and camels belonging to the hardy inhabitants of the region (PHOTO 23.2). The

PHOTO 23.1   Fault scarp near Mille, showing columnar jointing in the basalt of the Stratoid Series, and weathering of the tops of the columns into fragments which eventually become rounded and may fall down the scarp face onto the road (*at base of photo*) (2011)

PHOTO 23.2 A common southern Afar scene. Goats scrounge a meagre living from the scant vegetation growing on the black basalt and thin veneer of sediment (2013)

faulting pattern is intricate, with faults paralleling the Red Sea, the Main Ethiopian Rift and the Gulf of Aden often criss-crossing each other. Although most of the fault scarps are a few metres to a few tens of metres high, some reach heights of several hundred metres. One, known as the Gamarri scarp, northwest of Lake Abhe, is over a kilometre high!

A digital elevation model of southern Afar is shown in Fig. 23.1. Although the scale is too small to show all the faults, the main structures stand out clearly. The most conspicuous of these are the great graben: broad, fault-bounded plains covered by thick layers of sediment. The Tendaho Graben, for example, is about 50 km wide and its sediment fill up to 1.6 km thick. The sediments were deposited beneath the waters of a lake, or series of lakes, that once occupied the graben and they became thicker as the graben floor subsided. The narrower Dobi Graben is occupied by a shallow, very salty lake (currently being exploited for salt extraction) and a salt pan, remnants of a larger lake that filled it about 2000 years ago (PHOTO 23.3). Not all of the graben are named on Fig. 23.1, but it is easy to pick out where they are.

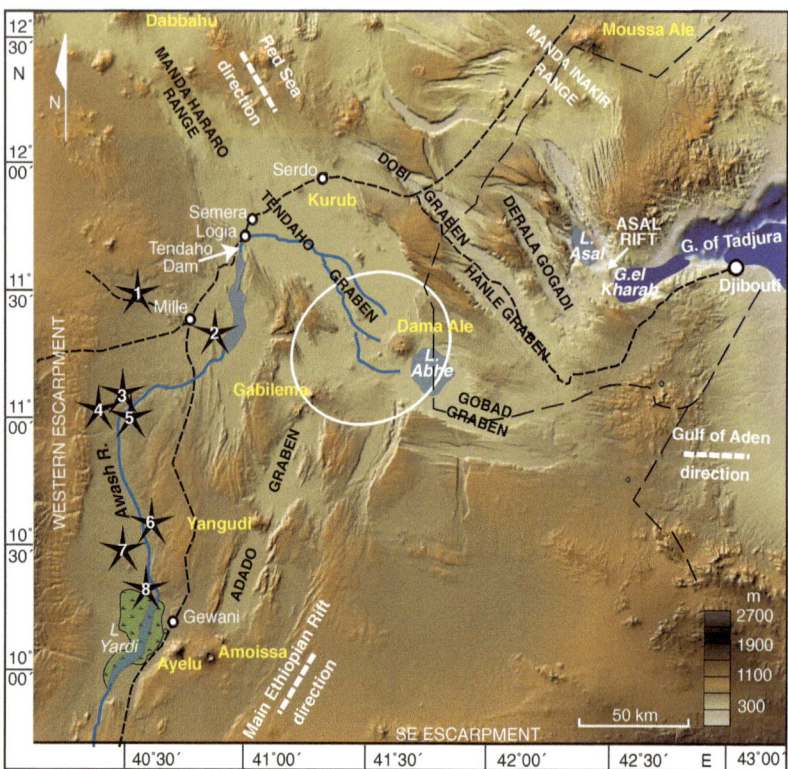

Fig. 23.1 Digital elevation model showing part of southern Afar. The *white ellipse* indicates the probable location of the Afar Triple Junction, and the *numbered black stars* indicate the main palaeo-anthropological sites of the middle Awash valley, listed in Table 23.1. DEM from GeoMapApp

Flows of the Stratoid Series almost certainly extend through northern Afar but are buried beneath the more recent products of the volcanic ranges. However, although by far the most widespread formation in Afar (they cover about two-thirds of it), the Stratoid Series has not been studied as closely as the more dramatic spreading ridges further north and remains as something of a puzzle. Seismic studies (examining the paths of earthquake waves through the rock) and gravity measurements have indicated that the total thickness of the Series is about 1500 m, but its base is not exposed anywhere and no-one knows for certain what lies beneath it. As is usual with geological puzzles, there is no shortage of theories. One is that the

PHOTO 23.3 The salt-covered floor of the Dobi Graben running across the centre of the photo, bounded by basalt ridges of the Afar Stratoid Series, here tilting toward the graben (2011)

Stratoid Series is underlain by continental crust which has stretched and thinned like a piece of pulled-out elastic as Arabia and Africa move apart. Certainly there is thin continental crust around the margins of Afar, close to the base of the escarpments, but there is no evidence that it underlies the whole of it.

A more widely accepted theory is that the Afar floor is formed entirely of basalts similar to those of the volcanic ranges. However, there are no distinct spreading axes over most of the region covered by the Stratoid Series. The spreading axes of southern and central Afar are located well to the east [the Manda Inakir range and the Asal Rift (see Fig. 21.1)], and to the western side is the Manda Hararo range. The broad expanse of Stratoid Series between them appears to have formed from lava which rose through many dispersed fissures before becoming confined to narrow spreading axes.

Whatever might underlie the Stratoid Series, two things are certain. Firstly, the total thickness of the earth's crust beneath Afar (about 20 km and becoming thinner toward the north), is less than half that beneath the highlands (about 45 km).

Clearly, the Afar crust is being stretched and thinned. Secondly, seismic studies (this time measuring the travel speed of earthquake waves) have shown that the mantle beneath the whole of Afar is hotter, and contains more molten material, than normal mantle. In fact Afar is underlain by the hottest mantle in the world! It is definitely sitting upon a very hot hot-spot.

23.2   The Afar Triple Junction

We can now take a closer look at where the Afar Triple Junction, the meeting point of the Red Sea and Gulf of Aden spreading axes and the Main Ethiopian Rift, might be located. On Fig. 23.1 the three directions can easily be distinguished. The Adado Graben with its bordering faults and line of silicic volcanoes (PHOTO 23.4) is really an extension of the Main Ethiopian Rift's Wonji Fault Belt, protruding into Afar. The Tendaho and Dobi Graben run parallel to the Red Sea. The Gulf of Aden direction is less clearly defined within Afar, as its spreading axis breaks up and curves around as it penetrates the continent, but faults representing it can be seen running WNW in the region of the Gobad Graben. Because each of the three fault systems consists of a broad belt of faults they don't meet at a single point, but you can see on Fig. 23.1 that they converge toward the region enclosed by the white ellipse in the southeastern part of the Tendaho Graben. The Main Ethiopian Rift faults actually stop very abruptly at the western edge of this graben (PHOTO 23.5). The white ellipse indicates roughly where the triple junction is at the moment, but it does not remain stationary. Because its three arms are widening in different directions and at different rates, and even all formed at different times,

PHOTO 23.4 Ayelu volcano near Gewani, a continuation of the line of silicic volcanoes of the Main Ethiopian Rift (2012)

PHOTO 23.5 Faults of the Main Ethiopian Rift (*right of photo*) abutting against those of the Red Sea trend (*left of photo*), between Mille and Logia. The photo was taken looking toward the northeast (2011)

it is continually on the move. Currently it is moving north-northeastward at about 1 cm per year.

It is interesting to note that the Tendaho Dam, completed in 2014 across the Awash River to provide irrigation for the Tendaho Sugar Plantation, is located just where the faults of the Main Ethiopian Rift meet those of the Red Sea trend, and very close to the triple junction itself. Perhaps not the best location for a dam—so near to where three plates are splitting apart!

23.3   The Tendaho Graben

The Tendaho Graben is one of the most important geological features in southern Afar. To begin with, the triple junction is located within it. Secondly, you can see from Fig. 23.1 that it is a continuation of the recently active Manda-Hararo volcanic range, and the cracks and volcanic activity that commenced in 2005 and were described in Chap. 22 are heading its way. An extensive field of hotsprings and fumaroles within the graben indicates that hot material is not far below the surface, and geophysical investigations using measurement of the electric conductivity of the underlying rocks have confirmed this, revealing that rock at a depth of about 7 km is so hot as to be partly melted. It is not surprising that this area is currently being investigated as a potential source of geothermal energy. At the southern end of the graben is a large volcano, Dama Ale, which last erupted in 1631, and prominent near the road between Logia and Serdo is a smaller basaltic volcano, Kurub, which has had no eruptions since it first formed about 300,000 years ago.

A sharp reminder that faults close to the margin of the graben are still active came in 1969 when a magnitude 6 earthquake destroyed the town of Serdo located on a ridge at its eastern side. We will visit Serdo in Chap. 24. Its broken buildings bear witness to the fact that the Tendaho Graben is in a particularly active region of Afar and, as we will see in Chap. 26, is a candidate for becoming the future Red Sea!

23.4   Spreading Axes of Central and Southern Afar

At the western end of the Gulf of Tadjura the Gulf of Aden spreading axis appears to terminate abruptly, at a small bulge called Ghoubbet el Kharab. However, like the Red Sea spreading axis, it doesn't actually terminate but moves into Afar. Unlike the Red Sea axis, though, it does not jump sideways but changes direction by about 90° and connects to a zone of faults, fissures and very fresh basalts extending to Lake Asal and known as the Asal Rift. As the name implies, rifting and subsidence are important here as well as volcanic activity. Further north, and a little offset to the east, is a longer, broader and more elevated zone known as the Manda Inakir Range. The Asal Rift and the Manda Inakir Range are similar to the northern spreading axes in that they are characterised by faults, open fissures and very fresh fissure basalts, but differ in not having silicic volcanic products. Both have been active in recent times. In the late 1920s a large fracture opened in the Manda Inakir Range and basalt erupted from a newly-formed cinder cone, and in 1978 a fissure-vent erupted in the Asal Rift, producing basalt lava flows for a week. These two zones of activity represent the complex extension into Afar of the Gulf of Aden spreading axis.

The volcanic rocks shown in pink and dark green on Fig. 21.1, to the east of these spreading axes, are older than the Stratoid Series and are quite different from them in composition. They include rhyolites, ignimbrites and basalts which, together with the silicic volcano Moussa Ale, are characteristic of a continental rift valley such as the Main Ethiopian Rift. Together with the Bab el Mandeb strait, they are the remnant of a continental rift which ceased to develop further when the Red Sea spreading axis, instead of continuing to grow along it, jumped westwards into Afar.

23.5   The Birthplace of Mankind

During the late 1960s a young French geologist was mapping the Awash River basin as part of his doctoral thesis research. His project area, close to the escarpment which marks the western margin of Afar, is thickly covered by sediments which have washed down from the highlands over the millennia since Afar first formed, mixed with layers of ash from eruptions of the southern Afar volcanoes and sediments deposited in ancient lakes. In the gullies around a dry river bed named Hadar (Fig. 23.1 (site 3) and PHOTO 23.6), he noticed such an abundance of fossil mammal bones: hippo, elephant, rhinoceros and many smaller creatures, that he drew the attention of palaeontologists (fossil experts) to the site. Over two three-month field seasons they sifted through the sand, silt and ash, unearthing more and more of this fossil treasure trove before coming upon the greatest find of all—the skeletal remains of a young female, neither ape nor human but something in between. The ages of volcanic ash layers interbedded with the

PHOTO 23.6 Beds of lake sediments and volcanic ash at Hadar, where the famous fossil "Lucy" (*Australopithecus afarensis*) was found in 1974. Today a plaque marks the spot where she was found. A group of vehicles at *right* indicates the scale (2013)

PHOTO 23.7 "Lucy", shortly after her discovery in 1974. She now lives safely behind the scenes in the National Museum in Addis Ababa. Only a plaster replica is on display to visitors (1974)

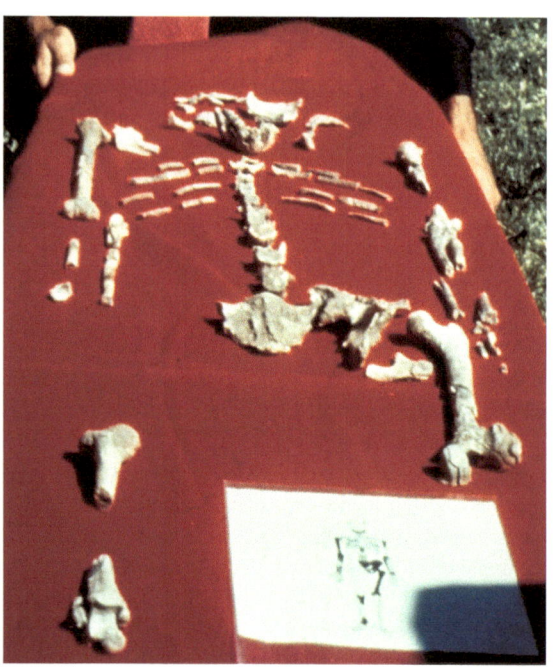

sediment in which she was found showed that she had lived 3.2 million years ago (PHOTO 23.7).[1] The American contingent of the team called her Lucy, after a Beatles song that happened to be playing at the time of the discovery, but the Ethiopians have christened her Dinkenesh ("you are wonderful"). Scientifically her name is *Australopithecus afarensis* meaning "the southern ape of Afar", though much about her is more human than ape-like.

Even though older *Australopithecus afarensis* fossils have since been found, Lucy has remained by far the most famous. Her discovery set off a kind of palaeo-anthropological stampede into southern Afar. Palaeo-anthropologists are specialists in the study of ancient humans and their predecessors, and clearly this region had something special to offer them. Numerous teams of various nationalities descended upon the middle reaches of the Awash River, vying with each other to find a new species or the oldest fossil, and to fill in gaps in the evolutionary

---

[1]Lucy's age was not firmly established until 20 years after she was found, as dating methods available at the time were insufficiently precise.

history of mankind. Many important discoveries have been, and still are being, made in this Middle Awash region. To describe all the different finds and their significance would take several books, and in any case I am not competent to venture into the complex field of human evolution. I have therefore indicated the main sites on Fig. 23.1, and summarised some of the important discoveries in Table 23.1 together with a key to help make sense of the rather formidable species names. Note that there is a bewildering plethora of names for pre-human species; the table and key show only the most significant ones that have been found in Afar.[2]

Some of the big questions that palaeo-anthropologists try to answer are: "When did apes become human (and more fundamentally, how do we define "human"?)?"; "When did they begin living on the ground rather than in trees?"; "When did they begin walking on two legs instead of four?"; "When did they start eating meat?"; "When were stone tools first used?"; "When did language originate?"... and many more. If you look carefully at the table you can see that the discoveries in Afar have been able to answer some of these questions.

A very big question is: "Why did humans evolve just here?" For it appears that they did, and that "the birthplace of mankind" is no exaggeration. Although apes, both fossil and living, are found in many parts of the world, and humans have spread almost everywhere, the intermediate steps such as *Ardipithecus ramidus*, *Australopithecus afarensis* and *Australopithecus garhi* are found only in east Africa. and most abundantly in the Middle Awash region of southern Afar. Of course one possible explanation is that they do exist elsewhere but have not been found. Another is that Afar offered an environment particularly favourable for the evolution of new and more advanced species.

Certainly, at the time these pre-human species lived, Afar was a very different place from the dry, inhospitable desert that it is today. The sediments which today occupy valleys between the volcanic ridges indicate the presence of once-substantial lakes and streams. The sandy plains around Afar's western and southern margins were laid down by rivers which today are dry wadis, carrying water only occasionally when there is heavy rainfall in the highlands. The fossil bones of large animals such as elephant, hippo and rhinoceros around the Middle Awash sites indicate that there was once plenty of water to drink and vegetation to

---

[2]New fossil finds in the Middle Awash region are being made at a rate too rapid to keep up with. As this book was about to go to press a 3.3–3.5 million year old jawbone was found at the Woranso-Mille site and assigned to a new species, *Australopithecus deyiremeda*. By the time you read the book, there will doubtless have been more discoveries.

Table 23.1 The main palaeo-anthropological sites in the middle Awash region

	Site	Major discoveries
1	Woranso-Mille	*Australipithecus afarensis* dated at about 3.4 My[a]
		Partial foot, possibly a transition between 4-legged and 2-legged mode of movement, also dated at about 3.4 My but different from *Australopithecus afarensis*
2	Ledi-Geraru	Jawbone of oldest known member of genus *Homo* dated at about 2.8 My
3	Hadar	First discovery of *Australopithecus afarensis:* "Lucy", a young female dated at about 3.2 My
		The "first family"—a large number of fossil bones similar to those of Lucy representing about 17 men, women and children
4	Gona	Oldest known stone tools dated about 2.6 My. Cut-marks found on animal bones are early evidence for butchery of large animals
		Ardipithecus ramidus dated about 4.4 My
		Homo erectus dated about 1 My
5	Dikika	Nearly complete skeleton of an *Australopithecus afarensis* 3-year old child, dated about 3.3 My (about 120,000 years older than "Lucy"). Nicknamed "Selam"
6	Bodo	600,000 year old cranium, intermediate between *Homo erectus* and *Homo sapiens.* Evidence of the skull having been de-fleshed after death (earliest indication of funereal ritual)
7	Aramis	First discovery of *Ardipithecus ramidus*, dated 4.4 My. Structure of foot and pelvis suggested ability to move on both 4 and 2 legs and to climb trees
8	Herto-Bouri	*Australopithecus garhi* dated at 2.5 My. Cut-marked animal bones found at site are early evidence for butchery of large animals
		Homo erectus dated about 1 My
		Early *Homo sapiens*, 2 adults and 1 child dated about 160,000 years

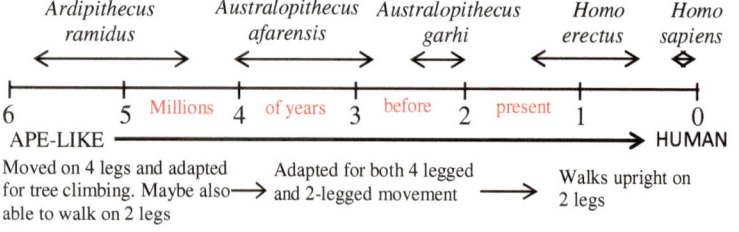

The numbers correspond to those on Fig. 23.1. Many discoveries have been made at each of these sites, and more are being made all the time. Only a few examples are shown in the table. The key below the table gives a brief explanation of the fossil species mentioned in the table
[a]My stands for "millions of years before present"

feed on. Fossil pollens tell of a forest environment, lush and well-watered, punctuated by grassy plains. All in all, it would have been a good place to live.

On the other hand, many other places in the world would have offered an equally agreeable living environment. The difference is that Afar also presented challenges which could stimulate adaptation. The times at which these pre-humans lived were also times of geological and climatic upheavals. Afar itself was still forming, faults were moving, accompanied by earthquakes, and volcanic eruptions occasionally coated the land with ash, killing vegetation and altering water courses. Earth movements sometimes tilted the land, causing some lakes to empty and others to fill. In addition the climate was changing. It was becoming drier, the forests were gradually dwindling and open plains taking over. The response of the pre-humans to their continually changing environment was to adapt, for example by developing limbs more appropriate for walking instead of climbing as the forests disappeared. Of course this took many generations to happen, but gradually those who did not adapt died out while those who did lived on as a new species in tune with the new environment. Only when conditions became too severe for adaptation were these early people forced to move away to find more hospitable climes, gradually spreading through the rest of the world.

Not only had this once been a good place to live; it is also a good place for fossil remains to be preserved and then to be found. Afar's climate, which has been hot and dry since those early dwellers departed, is ideal for preserving bones, pollen spores and other evidence needed for piecing together ancient species and their environment. It even helps with their excavation. During the highlands' rainy season water flows down into the wadis, washing out their banks and revealing the fossils they have been holding. Every year when the research teams return to their sites, new discoveries await them, freshly washed out of the wadi banks. Lastly, the sediments themselves provide a geological clock which enables the age of the fossils to be determined. The layers of sand and clay which contain the fossils are in many cases interbedded with layers of volcanic ash from the frequent eruptions of the silica-rich volcanoes in the region. The age of volcanic ash can be determined quite precisely, from measurements of the radioactive decay of potassium contained within its glassy fragments or by matching it with another volcanic ash of known age. It is hard to imagine a happier place for palaeo-anthropologists, apart perhaps from the discomforts of working in the desert heat, and the frequently bitter controversies between the different groups.

Earthquakes in Ethiopia

24

Considering her location at a point where three tectonic plates are breaking away from each other, it is no surprise that Ethiopia is a land prone to earthquakes. A full discussion of Ethiopian earthquakes is far beyond the scope of this book, and the reader is referred to the meticulously researched and comprehensive review, "Earthquake History of Ethiopia and the Horn of Africa", by Fr Pierre Gouin (see references at the end of this book) for further information. This chapter will merely brush the surface. To set the scene, we will return to Afar and visit the village of Serdo, situated almost exactly in its centre.

24.1 Serdo, March 1969

You may remember from Chap. 23 that Serdo is located near the eastern margin of the Tendaho Graben. Today it is little more than a cluster of huts, rubble and tumbledown stone buildings, but before 1969 it was an important regional centre. It contained a number of substantial buildings including the Police Headquarters, the Imperial Highway Authority building, a stone water tower, a hotel, a Telecommunications Office and the residence of His Excellency Bitwoded Ali Mirah, Sultan of the Aussa district of Afar. The town had a population of about 420, some of whom lived in stone or concrete houses, others in houses built of sticks with a roof of corrugated iron. Many people had covered the roofs of their homes with earth or stones as protection against the burning heat and desert winds.

At noon on Saturday March 28th 1969, most of Serdo's residents were indoors sheltering from the midday heat. Suddenly a sharp earth tremor shook the town.

© Springer International Publishing Switzerland 2016
F.M. Williams, *Understanding Ethiopia*, GeoGuide,
DOI 10.1007/978-3-319-02180-5_24

The people were alarmed but not too surprised—tremors were occasionally felt in this region, but this one was strong enough to make them rush out of their houses. However, no damage was done and people returned indoors to resume their siesta. But 2 hours later a roaring noise was heard, "like a great truck rushing towards us from the west" according to one witness—and the earth shook so violently that people, animals, furniture and walls were thrown eastwards, masonry buildings crumbled and homes with earth- or stone-covered roofs collapsed, crushing their occupants (PHOTOS 24.1 and 24.2). The school, a stone building with a thick concrete roof and at that time the only school in Afar, was flattened (PHOTO 24.3). The one piece of good fortune that day was that the earthquake struck after school was over; had it happened just a few hours earlier some 30 children would have been crushed to death.

The earthquake was recorded at seismic stations in Addis Ababa and Nairobi, and as far away as Uppsala in Sweden. Its magnitude was determined to be 6, and its epicentre, the point on the ground surface directly above the focus, or actual point of the dislocation which caused the earthquake, was located near the southwestern side of the town. It was followed by 250 aftershocks which continued

PHOTO 24.1 Aftermath of the 1969 Serdo earthquake. This photo was taken two weeks after the first shock struck the town (1969)

PHOTO 24.2   Another photo taken two weeks after the first shock struck Serdo. Notice on this photograph, and on PHOTO 24.1, that the structures made of sticks, wood, or corrugated iron remained standing while those of bricks, concrete and with heavy roofs collapsed (1969)

to shake the region for a further two months. One of these aftershocks was of magnitude 6.2, even bigger than the first shock, but it caused no further damage since everything had already been destroyed. The earthquake appears to have been caused by sideways movement along a fault running in a NE to SW direction (PHOTO 24.4), parallel to the direction of sea-floor spreading in the Red Sea and Gulf of Aden (see Fig. 22.1).

The casualty toll from the Serdo earthquake was 39 dead and 152 injured. Compared to earthquakes which make headlines in today's newspapers this seems trivial, and probably news of the disaster did not even reach beyond the borders of Ethiopia. The reason that the casualty rate was so low was of course that the earthquake occurred virtually in the middle of nowhere. It was only because it struck one of the very few towns in the region that there were casualties at all. Elsewhere there would have been no problem, as the traditional dwellings of the semi-nomadic Afar people are wisely constructed of sticks and matting and are impervious to earthquake damage.

PHOTO 24.3 Serdo's school, which had a heavy cement-slab roof, was completely flattened by the earthquake. Fortunately the students were not in school at the time. A new school, more appropriately constructed, has since been built on the same site (1969). Photo modified from Gouin (1979) and used by permission of the International Development Research Centre, Ottawa, Canada

If an earthquake of that magnitude occurred today beneath, or even close to, one of Ethiopia's ever-growing urban centres, it would be a very different story. For such an earthquake to strike Addis Ababa, for example, now a burgeoning metropolis of some 9 million residents and boasting a city-scape of high-rise buildings, would be a disaster of almost unimaginable proportions. There are now so many large towns in Ethiopia that the possibility of an earthquake affecting one or more of them is ever increasing. Not least among these is Semera, the modern administrative capital of Afar, located only 38 km from Serdo and, as you may remember from Chap. 22, right on the line of the recently active Manda Hararo Range.

PHOTO 24.4 Serdo's concrete water tower was tilted about 5° to the southwest by the earthquake, and much of its brick facing fell off. It remained standing, still tilted at this angle, until at least 2013. I have heard that it has since been demolished (2011)

24.2   Other Major Earthquakes in Ethiopia

The Serdo earthquake is by no means an isolated incident. Devastating earthquakes had occurred in Ethiopia on previous occasions. In 1842 an earthquake destroyed the town of Ankober, located on the western margin of Afar (Fig. 24.1). Ankober was at that time capital of the important province of Shoa, and a rival to Gonder as the capital of Ethiopia. The events were described by Major Cornwallis Harris, whom we have already met on his travels in the northern rift valley in Chap. 19: "....*The consequences were appalling. The soil, saturated with moisture slipped from the steep rugged slopes and rocks, heaved from their resting places, persued* (sic) *a sweeping course of devastation to the glens below...the destruction of life and property was altogether immense, and although shocks had often before been experienced, a similar calamity to the present had not befallen the country within the memory of man*". According to this description, most of the destruction was due not so much to the earthquake itself, as it was in the case of Serdo, but to landslides and rockfalls triggered by it. The situation was exacerbated by heavy rainfall which had saturated the soil and turned it to a muddy, slippery mass. Because seismic stations did not exist at that time, the epicentre and magnitude of the earthquake could not be determined, but it would have been located not far from Ankober and was probably of magnitude comparable to the Serdo 'quake.

On May 29th 1960 an earthquake of magnitude 6.6 shook the western escarpment again, near the town of Karakore some 100 km north of Ankober. Amazingly there were no casualties, but there was a huge amount of damage. A nearby village was completely destroyed, all masonry houses in Karakore collapsed, the main road cracked and was blocked by falling boulders, fissures opened, faults up to 2 m high formed and debris was dislodged from the hillsides (PHOTO 24.5). This earthquake shook and even damaged buildings as far away as Addis Ababa, and tremors were felt at Asseb on the Red Sea coast. As at Serdo, the Karakore earthquake was

PHOTO 24.5   This block of loose hillslope debris (the *brown material* at *left* of the photo), on the eastern side of the Borkenna Graben about 50 km south of Karakore, was probably dislodged from above during the 1961 earthquake (2011)

followed by many aftershocks—about 3500 altogether!—lasting altogether four months, and the buildings which survived the 'quake were the simple huts made of sticks, chika (mud and straw) and thatch.

The largest earthquake ever recorded in Ethiopia occurred in 1906 and had a magnitude of 6.8. Its epicentre could be located only very approximately from records received at distant seismic stations (at that time there were none nearby), and was estimated to have been near the foot of the western escarpment of the Main Ethiopian Rift, in the region east of Butajira. The local Gurage people thought that it marked the end of the world. Although there were no reports of casualties, the tremors were reported to be "exceptionally strong" in Addis Ababa, about 100 km away. At that time, only 10 years after it had been established as Ethiopia's capital city, there were no high-rise buildings of stone and concrete and very little damage occurred. Such an event would have very different repercussions if it happened today. One interesting consequence of this earthquake was a 25-m-high geyser which started up on a small island at the northern end of Lake Langano. This remained active for a number of years, but at the present time it has dwindled to a small surge of boiling water.

24.3 Distribution of Earthquakes in Ethiopia

The incidents described above are just a few examples of major earthquakes in Ethiopia. The "Earthquake History of Ethiopia and the Horn of Africa", mentioned above, lists more than 140 recorded between 1400 and 1977, not including aftershocks resulting from the same event. And of course many earthquakes are missing from the list. Prior to the 15th century written records of Ethiopian history are sparse, and there are none of earthquakes though they would certainly have been happening. From 1430 onwards there are a few scattered mentions, in Ethiopian and Arabic manuscripts, of earthquakes in the region, and from the beginning of the 19th century more reports emerged as European travellers began penetrating the country and producing detailed travelogues. Of course only the largest and most obvious events were reported. By the early 20th century seismic stations had been set up in several countries, including Russia and America, and it became possible to determine the magnitude and locate the epicentres of earthquakes, even from as far away as Ethiopia if the 'quakes were large enough. However, it was not until a seismic recording station was established in Addis Ababa in 1957, and by degrees further seismic stations in the region, that a systematic monitoring of all earthquakes, large and small, could be undertaken. At the present time eight permanent seismic stations are located in Ethiopia, and portable ones are always ready for emplacement should any unusual activity be recorded, as it was in the Dabbahu event of 2005 (see Chap. 22).

Figure 24.1 is a compilation showing the epicentres of earthquakes, of magnitude 4.0 and above, that have occurred in the Ethiopian region between 1900 and 2012. Epicentres of the older earthquakes are estimates based on the limited

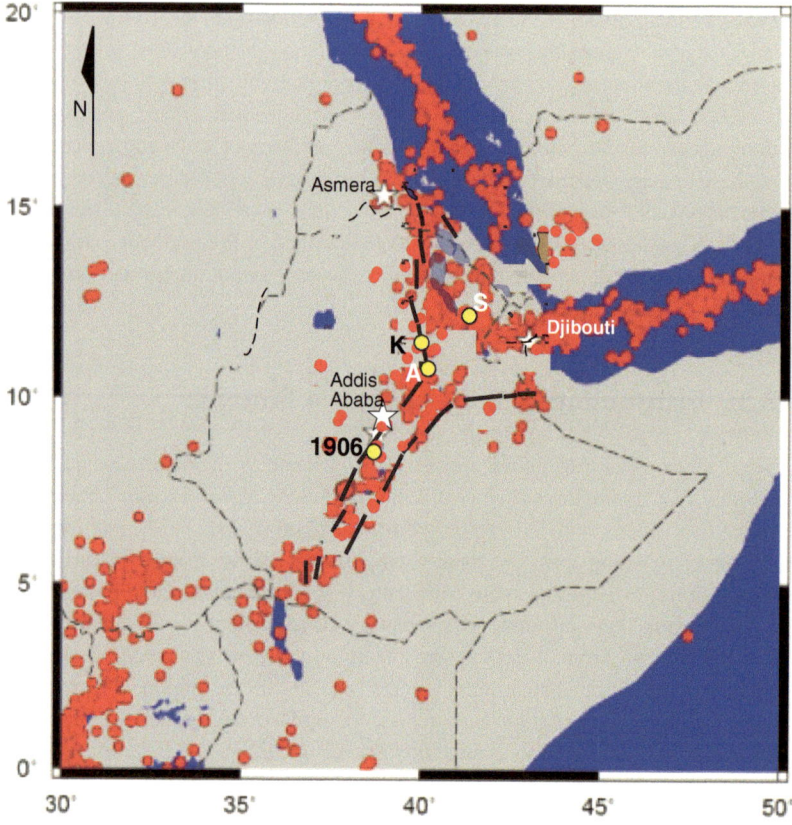

Fig. 24.1 Map showing epicentres of earthquakes (*red dots*) having magnitude 4.0 and greater in the Ethiopia region, between 1900 and 2012. The main rift and Afar margin faults are shown in *heavy black lines*, and the volcanic ranges of Afar are overlaid in *transparent blue*. The locations of the four major earthquakes described in the text are shown by *yellow dots*: A Ankober, S Serdo, K Karkore. The 1906 earthquake was not precisely located, but is thought to be in the region indicated. Epicentre map courtesy Atalay Ayele

information available, and many are missing altogether because there is no report of them. The more recent events, however, are very precisely located.

There are a number of things to note in this plot:

(i)   Epicentres are clustered along the spreading axes of the Red Sea and the Gulf of Aden.

(ii) At about 16°N the Red Sea epicentres split into two groups, one of which continues for a short distance along its central axis while the other moves over to its western margin and continues along the volcanic ranges of northern Afar.

(iii) South of latitude 14°N there are no epicentres along the central axis of the Red Sea.

(iv) A belt of epicentres curves through central-eastern Afar, connecting the Gulf of Aden cluster with those located along the northern volcanic ranges.

(v)   There is a concentration of epicentres along the western escarpment, particularly where it curves around to become the western margin of the Main Ethiopian Rift.

(vi) In the Main Ethiopian Rift most, though not all, earthquakes occur within the rift, particularly along the WonjiFault Belt and its continuation into Afar. There are actually numerous tremors within the rift whose magnitude is too small to show on the figure.

(vii) There are no epicentres along the southern margin of Afar, except at its eastern end.

From your reading of the previous chapters you will be able to see the reasons behind most of these observations. However, even though the plot covers more than 100 years of observations, this is a mere moment of geological time (about one and a quarter seconds in the calendar analogy we used in Chap. 2), and only gives a snapshot of what is happening at present. We can only imagine the large earthquakes that would have accompanied the early faulting of the rift margins, for example, or those which occurred along the borders of the Red Sea as Arabia began to break away from Africa. The pattern of the distribution does however provide some pointers which, together with other geological observations, enable predictions to be put forward about future developments, as we will see in Chap. 26.

Putting It All Together

At various points in this book I have promised that our journey through Ethiopia and observation of her geological features will throw some light upon the questions of why and how a continent breaks apart. It is with some trepidation that I will try to honour this promise, and I must warn the reader that at least as many questions remain unanswered as are resolved.

25.1 What Makes a Continent Break Apart?

There is a great deal of evidence that continents have been breaking apart and joining together again ever since the very earliest continental crust formed, around 4.4 billion years ago. The formation and break-up of Rodinia may in fact be the sixth such episode. What causes these cycles of break-up and coming together is the leading question in a debate which is well beyond the scope of this book. Ethiopia has, however, provided a valuable contribution to the debate, in revealing just why and how one small piece of a continent, located near the edge of the African Plate, is in the process of breaking apart.

From the previous chapters we have seen that three things have been working together to cause the split between the African, Arabian and Somali plates: a tectonic force, a weak lithosphere and a very hot mantle. A tectonic force simply means a major force causing, or caused by, movements of the lithosphere. The tectonic force in this case results from the Indian Plate dragging the African Plate alongside it as it moves toward Eurasia, as we saw in Chap. 8. The lithosphere is weak because it was patched together during the East African Orogeny, and is

© Springer International Publishing Switzerland 2016
F.M. Williams, *Understanding Ethiopia*, GeoGuide,
DOI 10.1007/978-3-319-02180-5_25

therefore formed of contorted metamorphic rocks with shear zones, fracture belts and various directions of structural "grain" to facilitate breakage. The abnormally hot mantle is due to the Afar Plume bringing up hot material from deep within the earth. Although the plume is hypothetical we have seen some convincing evidence in support of it, in the upward bulge of the Afro-Arabian Dome, and the outpouring of molten lava that formed the Trap Series. And, of course, it explains why the mantle is so hot beneath Ethiopia.

Is it a coincidence that these three circumstances occurred in the same region, and at around the same time? Here we enter the realm of speculation and will probe no further. The fact remains that they did, and that as a result the African Plate is breaking apart.

25.2   How Does a Continent Break Apart?

The first event in the break-up of this northeastern corner of Africa was uplift of the region, accompanied by eruption of lava flows, as a mass of hot material reached the base of the lithosphere (Fig. 25.1(i)). In Ethiopia we see the evidence for this in the great elevation of her highlands, surmounted by their thick pile of Trap Series volcanics. The uplift resulted in the formation of a triple junction: three cracks, or rifts, radiating out more or less from the peak of the uplifted area. The three arms of the junction did not, however, develop at the same time or at the same rate. The Gulf of Aden arm is most advanced and has fully reached the stage of sea-floor spreading, part of the Red Sea has reached this stage, and the Main Ethiopian Rift, the least advanced arm, is still simply a crack in the continental crust. Afar, where the three arms almost, but not quite, connect contains volcanic ranges which are at an in-between stage.

From the three arms of the junction, and their near-meeting point in Afar, we can piece together the process by which a crack in the continent may develop into an ocean floor.

Beginning with the earliest stage, illustrated by the Main Ethiopian Rift, we have seen that the rift margins formed first by bending or flexing, and then by faulting with much accompanying volcanic activity. The faulting and volcanic activity gradually moved inward from the margins to a narrow zone within the rift itself (Fig. 25.1(ii)). They worked in a feed-back loop, magma lubricating the faults and easing their movement, and the faults enabling passage of magma to the surface. This is why earthquakes that occur within the rift, though numerous, are generally of small magnitude—the lubricating effect of the magma enables the faults to move less jerkily. This is the stage the Main Ethiopian Rift is at now. It is

Fig. 25.1 The main events
in the break-up of the
African-Arabian plate. The
Main Ethiopian Rift, Afar
and the Red Sea/Gulf of
Aden are at different stages
of the process of rifting and
separation

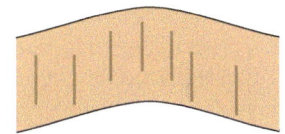

(i) Uplift and doming; eruption of flood lavas
through dykes (Ethiopian highlands)

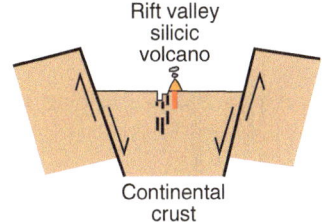

(ii) First stage of rifting:continental rift valley
(Main Ethiopian Rift)

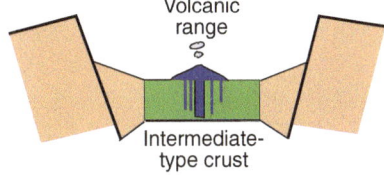

(iii) Intermediate stage (Afar)

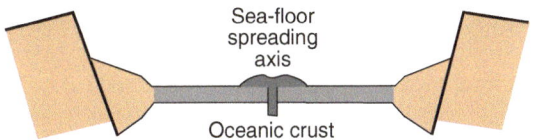

(iv) Final stage: sea-floor spreading (Gulf of Aden
and part of the Red Sea)

still entirely underlain by continental crust. The rate of widening is very slow
because the continental crust does not stretch easily and, although some fractures
and fissures have formed, it has not yet broken apart. The rift may or may not
develop beyond this stage.

The next stage, in between a continental rift and sea-floor spreading, is apparent in Afar, the only place in the world where it can be observed (Fig. 25.1(iii)). The crust in Afar is thinner than continental crust, but not as thin as oceanic crust. The composition of the Stratoid Series basalts which cover much of Afar, and of those being produced along the volcanic ranges, is not quite like that of continental basalt such as the Trap Series, neither is it quite oceanic. The volcanic activity in Afar was first spread over a broad region (the Stratoid Series), but has now become focussed along narrow bands (the volcanic ranges) which are almost, but not quite, sea-floor spreading axes. The split of the continent is nearing its final stage here—but has not quite reached it yet.

The final stage is sea-floor spreading, the creation of a new ocean floor and a complete break between the continental fragments (Fig. 25.1(iv)). The Gulf of Aden and part of the Red Sea are at this stage, having passed through the first two to reach it. It just remains for the corner of the Arabian Peninsula to detach itself from its Afar sticking-point—and the process will be complete.

25.3 Is the Ethiopian Situation a Typical Example?

Since the Ethiopian region is the only place where we can currently observe a continent in the various stages of breaking apart, it is difficult to say whether the combination of circumstances and the sequence of events that we see here are typical. The only comparison we can make is by attempting to reconstruct some of the break-up processes between other continental fragments of Gondwana, or those of the ancient northern supercontinent Laurasia. These happened so long ago, however, that sea-floor spreading has pushed the fragments far apart, the original rift margins have been submerged by the sea and within-rift features are deep beneath the ocean, so that evidence of the early events is hard to trace. What evidence there is, for example from the flood basalts of India and Madagascar, rift margin faults detected off the coast of South Australia, and the geology of the coasts of Greenland and Norway, separated by the mid-Atlantic Ridge and the Iceland hot-spot, is not inconsistent with the example from Ethiopia. Much depends upon the structure, rock types and past history of the continent concerned. In some cases the earlier stages in the break-up process, uplift and flood vol-canism, are not apparent and may not have occurred at all; in some the breakage does not appear to have taken place along any obvious line of weakness. Nature has provided too many variables for almost any geological process to be truly "typical".

Ethiopia has, however, provided us with a snapshot of a breaking continent. Comparing this snapshot to the other situations is rather like comparing a recent, clear photograph of a landscape with old ones showing different sections of it, and taken from various angles. The old photographs have been scratched, torn, even dropped in water, and are blurred with age. Although the new photograph may help to identify some of the features shown on the old ones, others are too obscure to interpret. In addition, the photographs showed different views in the first place. Nevertheless the new photo provides the best chance of interpreting the old ones and of working out the overall picture. This is the opportunity that our study of Ethiopia and her geology have given us.

26.1 The Long Term Future—Some Speculations

Figure 26.1(i) illustrates a hypothetical scenario 10 million years hence, based on the assumption that the plates continue moving as at present. The situation in the Afar region is shown in more detail in Fig. 26.1(ii). In this scenario the Red Sea and the Gulf of Aden have become wider and have finally connected with each other, in a zig-zag fashion, via the volcanic ranges of Afar which have developed into spreading axes. If you look back to Fig. 24.1 you can see that the earthquake epicentres in Afar more or less follow this zig-zag line, emphasising the fact that it is very active and mobile. The newly-developed spreading axes are offset sideways from each other, just as they were when they were volcanic ranges, and are connected by faults shown as thin black lines on Fig. 26.1(ii). Faults which connect sections of spreading axes are known as transform faults and movement along them is sideways rather than vertical. The block of continental crust of which the Danakil Alps are formed has become part of the Arabian Plate, and the Arabian and African Plates are now completely separated. Of course things may not happen exactly this way—it is just one possibility.

The Main Ethiopian Rift is shown as having widened to become an open sea. This is what would happen if it kept on widening at its present rate. Northwestern and southeastern Ethiopia would belong to separate continents. However, study of ancient rift valleys in other regions has indicated that when three spreading axes meet at a triple junction, as the Red Sea, Gulf of Aden and Main Ethiopian Rift do, two of them develop into true oceans while the third stops widening and stays the way it is. This is what may happen to the Main Ethiopian Rift and to the rest of the

© Springer International Publishing Switzerland 2016
F.M. Williams, *Understanding Ethiopia*, GeoGuide,
DOI 10.1007/978-3-319-02180-5_26

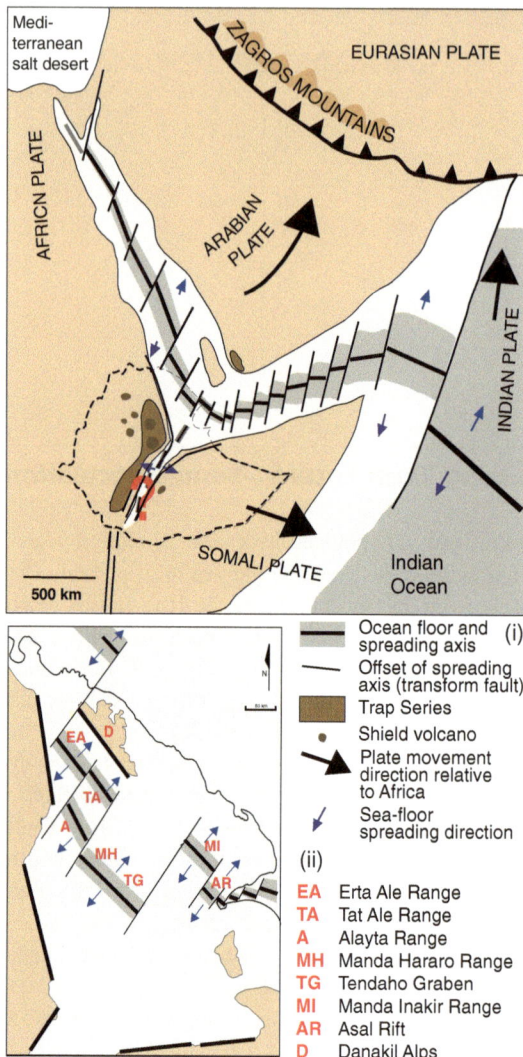

Fig. 26.1 (**i**) The possible situation 10 million years from now, if the plates keep on moving as they are at present (Adapted from Bosworth et al. 2005) and (**ii**) how the Red Sea and the Gulf of Aden may connect through Afar as the volcanic ranges develop into spreading axes

African Rift System to the south, which is why it is marked with a red question mark on Fig. 26.1(i). Ethiopia, apart from losing a bit of eastern Afar, may remain intact.

You will notice on Fig. 26.1(i) that the Mediterranean Sea has become a salt desert. This is because the general northward movement of Africa and Arabia has caused its western end, already very narrow at the Strait of Gibraltar, to close. It has become cut off from the ocean which replenished it, and its water has evaporated.[1] In addition, the Arabian Peninsula has completed its collision with Eurasia, and the Persian Gulf and Gulf of Oman have gone. The collision will have slowed down the movement of Arabia, and the whole process therefore may be grinding to a halt.

Many other changes will have taken place. Erosion will have removed much of the pile of Trap Series volcanics from the highland regions, leaving isolated ambas. The West Tana Escarpment will have been eroded back to the lake which, with nothing to hold it in, has emptied. The tributaries of the Tekeze River will have eaten far back into the Semien Mountains and the fragile sandstone mesas and pinnacles of the Mekele Basin and Tigray worn away. For those who like to speculate, there is much scope here!

26.2 The Near Future—Some Apprehension

Obviously we have no control over these long-term developments, and indeed it is doubtful whether the human species, as we know it, will be there to see them. The future in terms of millions of years is interesting to speculate upon, but there is little need to worry about it. What we can control, and what we do need to be concerned about, is what is happening at present and which will affect future generations. Of course it is all too easy to be nostalgic for the past, and to reminisce on how much better things were in the "old days", but sometimes I feel a real twinge of anxiety for the fate of some of the geological sites which I knew during my early days in Ethiopia and have since revisited. Some instances are the once surreal landscape of salt mounds and hotprings at Dallol which has, for whatever reason, been in a state of decline over recent years; the spectacular flow of black lava at Fantale which is being broken up for the passage of a new road and railway

---

[1]This has in fact happened on a previous occasion, between about 5 and 6 million years ago, due to a world-wide lowering of sea level, and the Mediterranean is already floored by up to 3 km of salt.

line; the magnificent Tis Isat Falls which have lost much of their grandeur since water was diverted from them for hydroelectricity.

Geological heritage is as much the essence of a country as are its cultural, historical and ecological heritages, but it is one that can easily be overlooked. Not only are people often unaware of it, but it comes with valuable resources which tempt exploitation: rocks to be quarried, minerals to be mined, rivers to be dammed and so forth. Certainly all of these resources are necessary and desirable, but it is also desirable that care be taken to extract and utilise them without damaging unique, significant or beautiful geological features. I must admit that when I see, for example, a perfect, symmetrical cinder cone being torn apart for road metal I feel a surge of sadness, and to see one of the unique volcanic blisters formed from the beautiful Fantale welded tuff being broken up for building stone, or used as the town rubbish dump, is heartbreaking.

Even though I think of Ethiopia as my second home, I am very much aware that she is not my country and that during my travels I am only there as a guest. It is not my place to provide opinions on how things should or should not be done. I hope however that in writing this book I have demonstrated that Ethiopia's geology is an invaluable resource in its own right, an asset for her people to be immensely proud of, and one which should be protected and treasured. It is unlike that anywhere else in the world, and is not only a fount of information to earth scientists in their quest for understanding how the earth works but also a source of wonder to anyone who has even a small understanding of it. It is the foundation of Ethiopia's magnificent and varied scenery which never fails to awe even the least geologically-inclined spectator, and draws visitors from around the globe to this very special country.

> Here is your country. Cherish these natural wonders, cherish the natural resources, cherish the history and romance as a sacred heritage, for your children and your children's children. Do not let selfish men or greedy interests skin your country of its beauty, its riches or its romance.
>
> —Theodore Roosevelt

Glossary

Note: The names of geological eras, periods and epochs are not included in this glossary. For those please refer to the geological time scale on page 13

Afro-Arabian Dome Uplifted region of Ethiopia, Eritrea and part of the Arabian peninsula, possibly caused by the impact of a **mantle plume** during late Mesozoic and Tertiary times

Agglomerate A rock made up of cemented fragments of volcanic rock 2 cm or more in diameter

Amazonite A type of potassium **feldspar** with a greenish-blue colour due to impurities in the crystal structure

Amba See **mesa**

Anhydrite A mineral composed of calcium sulphate. It is similar to **gypsum** except that it contains no water in its crystal structure

Arabian-Nubian Shield An area of Precambrian rocks covering much of the Arabian Peninsula and parts of Ethiopia. The rocks were formed during the **East African Orogeny**, by compression and **metamorphism** of what was originally ocean floor and **island arcs** between the colliding African and Indian **cratons**

Ash (volcanic) Very fine fragments, mainly glassy, produced in a volcanic eruption. Often grey and powdery in appearance

Asthenosphere The part of the earth's **mantle** between about 80 and 200 km beneath the earth's surface. It lies between the **lithosphere** and the lower mantle

Atrophic Refers to an environment, particularly lake water, that is deficient in oxygen

© Springer International Publishing Switzerland 2016
F.M. Williams, *Understanding Ethiopia*, GeoGuide,
DOI 10.1007/978-3-319-02180-5

Azania A small craton, corresponding very roughly to today's Madagascar, which was squeezed between Africa and India during the **East African Orogeny**

Basalt A dark, heavy, fine-grained volcanic rock consisting mainly of calcium **feldspar** and dark-coloured iron- and magnesium-rich minerals

Basement The oldest rocks in a region, upon which the later rock formations are built

Basin This term has two meanings: (i) a general term for an extensive low-lying area which has formed due to subsidence; (ii) the area which includes a river and all its tributaries (e.g. the Nile Basin). (See also **rifted basin**)

Beryl A silicate mineral formed of beryllium, aluminium and **silica**. It is frequently green in colour and may be used as a gemstone

Black smoker A vent on the sea floor which emits a cloud of black material consisting of hot water, gases and sulphur-bearing minerals

Blister (volcanic) A hollow mound formed by the pressure of gas trapped within a solidifying **lava** or **welded tuff**

Bomb (volcanic) A large blob of **lava** which has solidified before reaching the ground

Brachiopod A marine shellfish with hinged upper and lower shells. Brachiopods were abundant during the Palaeozoic and Mesozoic eras, and evolved through many different forms, enabling them to be used to estimate the age of the rock in which they are found

Caldera A large volcanic depression, usually several kilometres in diameter with steep inner walls, formed when the top of a volcano collapses into the space vacated by the material which has erupted from the volcano

Chert A cryptocrystalline (sub-microscopic) form of **silica**

Cinders Small **vesicular** fragments, black, red, brown or yellowish in colour, usually derived from **basalt** lava rich in gas

Cinder cone A small volcanic mound, usually conical and often containing a crater, formed of **cinders**

Cirque A large hollow formed by erosion at the head of a glacier

Clay Very fine sediment whose individual particles are less than 0.002 mm in diameter

Cleavage plane In metamorphic rocks, this refers to a plane along which the rock tends to break easily, due to the layered arrangement of platey minerals

Columnar jointing A type of jointing, most commonly occurring in **basalt**, that gives the rock **outcrop** the appearance of being made up of a series of closely packed, approximately hexagonal columns. This type of jointing is due to contraction of the rock mass as it cools after eruption

Continental crust See **crust**

Core The innermost part of the earth, formed of iron and nickel

Craton An old and very stable portion of continental lithosphere that has remained intact for over a billion years

Crust The outer layer of the earth, about 6 km thick beneath oceans (**oceanic crust**) and 35 km thick beneath continents (**continental crust**), formed of rocks which are less dense than the material of the underlying mantle

Deinotherium A prehistoric relative of modern-day elephants that existed between the Middle Miocene and Early Pleistocene

Diatom A microscopic, unicellular organism (alga) that has a porous cell wall made of **silica**

Diatomite A soft, white, powdery sedimentary rock formed by the accumulation of the siliceous remains of **diatoms**

Digital Elevation Model A 3-dimensional representation of the earth's surface compiled from a number of sources including radar and satellite imagery

Dolomite A sedimentary rock formed of calcium magnesium carbonate, a mineral which is also called dolomite

Dormant volcano A volcano which has ceased to erupt **lava** but which still shows signs, such as **fumaroles**, of recent and possible future activity

Dunite An **ultramafic** rock consisting of at least 90 % of the mineral **olivine**. It is a major constituent of the upper part of the earth's **mantle**

Dyke A narrow **igneous** intrusion that has penetrated a crack in a pre-exisitng rock. Dykes often represent the feeders for **lava** which has erupted onto the surface

East African Orogen A belt of metamorphic and igneous rocks which extends along the western side of the Arabian Peninsula and the eastern side of Africa as far as Mozambique. From here it extends into western Antarctica. It was formed during the **East African Orogeny** and consists of two distinct sections: the **Arabian-Nubian Shield** in the north, and the **Mozambique Belt** in the centre and south

East African Orogeny A major mountain-building event resulting from the collision of Africa, **Azania** and India during the late Precambrian era. The resulting mountain belt, now worn down to its roots, is known as the **East African Orogen**

En échelon Term applied to **faults**, fractures or **rifts** which are displaced from each other in a sideways manner

Epicentre The point on the Earth's surface that is directly above the focus (point of origin) of an earthquake

Erosion Removal and transportation of **weathered** material, mainly soil and rock debris, by natural agencies such as wind, water or ice

Erosional remnant A landscape feature which is left standing after the material surrounding it has been eroded away

Erosional scarp A steep slope or cliff formed as a result of **erosion**, particularly **headward erosion** by rivers (see also **escarpment**)

Escarpment or scarp A steep slope or cliff formed by vertical movement of the earth's crust along a fault (**fault scarp**) or by erosion (**erosional scarp**)

Explosion crater See **maar**

Extrusive rocks Rocks formed by solidification of **lava**: **magma** which has reached the earth's suface. Because the lava has cooled rapidly, extrusive rocks are usually fine-grained. They are often referred to as **volcanic rocks** as they are frequently erupted from a volcano. (See also **igneous rocks**)

Fairy chimney See **hoodoo**

Fault A fracture in rock along which there has been an observable amount of displacement either vertically or horizontally

Fault scarp A topographic **escarpment** formed as a result of vertical displacement on a **fault**

Feldspar Feldspars comprise the commonest group of rock-forming minerals. They are composed of sodium, potassium, calcium or a combination of these, together with silicon and oxygen

Fiamme Black glassy streaks in **welded tuff**, formed by rapid solidification of large liquid drops

Flexure A situation where rock layers have bent over, rather than breaking to form a **fault**

Flood lava or flood basalt Lava (generally **basalt**) which occurs as extensive horizontal layers. The lava has flowed out from fissures or vents rather like flooding water would. Each layer represents an individual flow

Formation A group of rock layers, or strata, of similar age, which consist of similar rock types and were deposited under a similar set of conditions. The term is also used in a different sense to mean a shape or structure e.g. the pillar-like formations at "New York"

Fossil Remains or trace of a living organism, preserved in rock

Fumarole A small vent from which vapours (mainly steam) and hot volcanic gases are emitted

Gabbro A coarse-grained igneous rock with composition similar to that of basalt

Gastropods A group of marine and terrestrial animals including snails which have a soft body and generally a single shell. They have existed from the Late Cambrian to the present time and, like brachiopods, evolved through a large variety of forms which enable the ages of rocks containing them to be distinguished

Geothermal Descriptive of heat coming from within the earth. Geothermal manifestations included hotsprings, **geysers** and **fumaroles**

Geyser A hotspring that intermittently ejects a column of water and steam into the air

Gneiss A metamorphic rock showing distinct foliation (layering), representing alternating layers composed of different minerals

Gondwana A landmass ("supercontinent") that existed from about 530 to 180 million years ago and consisted of present-day South America, Africa, India, Madagascar, Antarctica and Australia

Graben A valley formed where a strip of the earth's crust has dropped between parallel **faults**. The terms "graben" and "**rift valley**" are sometimes used synonymously, but generally "graben" is used to refer to a smaller-scale feature

Granite A coarse-grained intrusive igneous rock formed mainly of potassium **feldspar** and **quartz**, with minor amounts of dark minerals and calcium feldspar

Granitoid General term for any coarse-grained intrusive **igneous rock**, not necessarily granite

Guyot or seamount An isolated, flat-topped underwater volcanic hill

Gypsum A mineral composed of hydrated calcium sulphate. Similar to **anhydrite** except that gypsum has water molecules in its crystal structure whereas anhydrite does not

Headward erosion Upstream cutting of a river into the land over which it flows

Hoodoo or fairy chimney A pillar of soft rock, often with a capping of harder rock that has protected it at least temporarily from erosion

Hornito A mound of congealed lava that emits vapours and occasionally small amounts of liquid lava

Horst A generally flat-topped ridge formed by uplift between two opposing normal faults (the reverse of a **graben**)

Hot-spot See **mantle plume**

Hyaloclastite ring A crater surrounded by a ring formed of glassy fragments, together with **tuff** and **ash**. Thought to be the result of an underwater eruption

Igneous rocks Rocks formed by the cooling and solidification of **magma**, either below the earth's surface (**intrusive igneous rocks** e.g. **granite**) or on the earth's surface (**extrusive igneous rocks** e.g. **basalt**)

Ignimbrite A rock formed as a result of a **nuée ardente** volcanic eruption. It consists of a mixture of crystals, rock fragments, glassy fragments and often pumice fragments. If the fragments are fused together it is termed a **welded tuff**, though the terms are often used synonymously

Inselberg (Literally "island mountain"). An isolated rocky hill rising from a surrounding plain

Intertrappean sediments Sediments Occurring between layers of the **Trap Series** volcanics, deposited by lakes, rivers or on land during a break in the sequence of eruptions

Intrusion A body of coarse- or medium-grained **igneous rock**, formed from **magma** which has pushed its way into pre-existing rock and solidified before reaching the surface

Intrusive rocks A general term for **igneous rocks** which have formed from **magma** which has cooled below the earth's surface. Because the cooling was slow, intrusive rocks are generally coarse-grained

Island arc An arc-shaped chain of volcanic islands, located in a zone where oceanic crust is being subducted beneath oceanic crust close to a continental margin

Joint A fracture in rock along which no movement has occurred. Joints are generally the result of cooling processes (as in **columnar jointing**) or release of pressure as material above the rock is removed by **erosion**

Karst A landscape resulting from the dissolution of limestone and characterised by features such as caves, sinkholes and underground streams. Karst features which are observed at the earth's surface, such as gullies, hollows and fluted rock surfaces are termed **surface karst**

Laurasia See **Pangea**

Lava Molten rock material which is erupted onto the earth's surface and solidifies to form volcanic rocks such as **basalt**, **rhyolite** and **trachyte**. The term is also used to refer to the rock formed by the solidification of this molten material

Lignite A sedimentary rock formed from compressed vegetation. It is combustible and is a very low-grade type of coal

Limestone A sedimentary rock formed of calcium carbonate, either by precipitation from solution in water or from compressed skeletal fragments of marine organisms (in which case it may be referred to as fossiliferous limestone)

Lineament A linear feature in the landscape, usually on a large scale, often representing an underlying structure such as a fault or other line of weakness resulting from a prior geological event

Lithosphere The outermost solid "skin" of the earth, consisting of the **crust** plus the uppermost **mantle**

Maar Also known as **explosion crater**. A volcanic feature, generally broad in comparison to it height, with a deep crater and a prominent rim built of layers of **tuff**. Maars are believe to results from the violent jetting of **lava**-heated steam which causes the explosive blow-out of rock fragments and lava droplets, which fall as **tuff** and **cinders** that form the crater rim

Mafic A term applied to rocks or minerals rich in **ma**gnesium and/or iron (**Fe**)

Magma Molten rock material before it has reached the earth's surface

Magma chamber The location of a large underground pool of molten or partially molten rock

Mantle The layer of the earth between the **core** and the **crust** (extending from about 30 km to about 2900 km beneath the earth's surface), formed of dense rock material rich in iron and magnesium. It consists of the **lower mantle**, which is very dense and rigid, the **asthenosphere** which is solid but slightly plastic and able to flow a little, and the **uppermost mantle** which is a thin, rigid layer closely linked to the crust together with which it forms the lithosphere

Mantle plume A hypothetical flow of very hot rock which rises from deep within the earth's mantle, possibly from the core-mantle boundary. It spreads out when it reaches the base of the lithosphere to produce a zone of abnormally hot and partially molten mantle, and is manifested at the earth's surface as a region of volcanic activity known as a **hot-spot**

Marginal graben A narrow faulted valley running along the margin of a **rift valley**

Mesa A flat-topped, steep-sided hill. In Ethiopia mesas are known as **ambas**

Metamorphism The process by which the minerals and/or texture of a rock change as a result of being subjected to heat and/or pressure, and/or the action of chemically active fluids

Metamorphic rocks Rocks formed as a result of **metamorphism**

Mica A group of flakey silicate minerals, often found in **granites** and **pegmatites**

Mid-ocean ridge A ridge along a plate boundary where oceanic crust is separating from oceanic crust. As the name implies, the ridge usually occurs along the centre line of an ocean. It consists of volcanic rock which has erupted from vents and fissures to fill the gap between the separating plates, a process known as **sea-floor spreading**. (See also **spreading axis**)

Migmatite A metamorphic rock that has partially melted, and the molten material has then re-solidified. It is called migmatite ("mixed rock") because it is really a mixture of a metamorphic and an igneous rock

Moraine Accumulation of rock debris deposited by a glacier either along its sides (lateral moraine), at its snout (terminal moraine) or along its floor (ground moraine)

Mozambique Belt A belt of metamorphic rocks that extends from western Antarctica through the eastern side of Africa to the **Arabian-Nubian Shield**. It formed as a result of the collision of Africa, Antarctica, Mozambique (**Azania**), and India during the **East African Orogeny**

Mudstone A very fine-grained sedimentary rock formed of compressed **clay** or mud

Nepheline A silicate mineral similar to a sodium **feldspar** but containing a lower proportion of silica

Nuée ardente A mixture of crystals, **lava** droplets and incandescent **ash** in a turbulent cloud of immensely hot gas, the result of a violently explosive volcanic eruption

Obsidian Volcanic glass, usually black, formed where lava, generally rich in silica, cools and solidifies so rapidly that there is no time for crystals to form

Oceanic crust See **crust**

Olivine A **mafic** silicate mineral containing iron and magnesium and generally green in colour

Opal A mineral formed of minute spheres of **silica** surrounded by water molecules. Used as a gemstone

Ophiolite An assemblage of mafic and ultramafic rocks, with some sedimentary material, the metamorphosed remains of a **subduction** zone and oceanic floor which have been squeezed upwards and outwards during a continental collision

Orogen A belt of metamorphosed rocks resulting from the collision of continents during an **orogeny**. Initially an orogen is a range of high mountains, later worn down to their roots by **weathering** and **erosion**

Orogeny A mountain building episode resulting from a collision of continents

Outcrop Rock which is exposed at the earth's surface

Palaeoanthropology The study of human evolution using evidence from **fossil** remains

Pangea A super-supercontinent that existed between about 300 and 180 million years ago, consisting of the combined supercontinents of **Gondwana** (the present southern continents plus India) and **Laurasia** (the present northern continents: Europe, Asia and North America)

Pegmatite An intrusive igneous rock consisting of grains that are more that about 2.5 cm in size

Pele's Hair Threads of volcanic glass formed when droplets of **lava** are thrown into the air and drawn out into long hair-like strands

Phonolite An extrusive igneous rock similar to trachyte but containing feldspar minerals which have a low content of silica. Frequently forms plugs and domes

Pitchstone A black, mostly glassy rock similar to obsidian, but because it has a higher water content it is less glassy in appearance and generally contains scattered white crystals

Plate (tectonic plate) A rigid segment of the **lithosphere** which may consist of oceanic lithosphere only, or both continental and oceanic lithosphere. Plates are separated from each other by **plate boundaries**, and move around on the surface of the earth, on top of the non-rigid **asthenosphere**

Plate boundary See **plate**

Plug, volcanic A steep-sided hill formed from **lava** which has solidified within a volcanic vent and is then exposed when the surrounding volcano is eroded away. In some cases the solidified lava is pushed upwards by new lava rising below it

Potash A general name for salts of potassium. These minerals are often mined for use as fertilizers

Pumice A pale, very light-weight porous rock, generally with a composition similar to that of **rhyolite**. It is formed of glassy threads surrounding **vesicles**, and resembles a foam in appearance

Pyroclastic Literally "fire broken". A general term referring to loose material or rocks formed as the result of an explosive volcanic eruption

Pyroxene A mineral consisting of iron, magnesium, calcium or a combination of these, combined with silicon and oxygen. An essential constituent of **basalt**

Pyroxenite An **ultramafic** rock consisting mainly of **pyroxene**

Quartz A common mineral formed of silicon and oxygen (silicon dioxide), often having a glassy or milky appearance

Radiometric dating Refers to a number of methods for dating rocks, based on radioactive decay of elements which are present in them

Retreat scarp An **escarpment** which retreats over time, generally due to **headward erosion** by rivers running from it

Rift See **rift valley**

Rift valley, often abbreviated to rift A strip of the earth's crust which has subsided between parallel **faults**

Rifted basin A section of the earth's crust which has subsided between **faults**. A more general term than **rift valley** and may refer to a broader feature

Rhyolite A fine-grained volcanic rock, usually light in colour, composed mainly of **quartz** and sodium or potassium **feldspar**. It is the fine-grained equivalent of **granite**

Rodinia A hypothetical supercontinent, comprising the ancient cores of all of today's continents, which is postulated to have existed from about 1100 to 850 million years ago

Sandstone A sedimentary rock formed of sand-sized (1/16 to 2 mm) rock or mineral fragments

Sapphire A gemstone, usually blue; a variety of the mineral corundum which is an oxide of aluminium

Scarp See **escarpment**

Scoracious Term applied to a **basaltic** rock containing many gas bubbles or **vesicles**

Scoria or scoriaceous basalt Basaltic rock fragments or blocks containing many **vesicles** (gas bubbles), but generally less vesicular and more massive than **cinders**

Sea-floor spreading The process by which ocean floors widen as material is erupted along a **mid-ocean ridge** or **spreading axis** to create new sea floor

Sediment Loose material such as sand, clay and gravel, which has been formed by breakdown of rocks by **weathering**, and transport and deposition of the broken-down material by **erosion**

Seismometer An instrument for detecting earthquake waves

Sedimentary rock A rock formed of **sediment** which has become hardened, either by compression or by the grains of sediment being cemented together. Some sedimentary rocks such as **limestone** are formed by chemical precipitation of material

Serpentine A group of minerals which are often green in colour and usually fibrous or platey, composed of iron, magnesium and hydroxyl (OH) ions. They are formed by metamorphism, in the presence of water, of ultramafic rocks

Serpentinite A rock formed mainly of **serpentine** minerals

Shale A **sedimentary rock** formed of clay or silt-sized particles (less than 0.063 mm in diameter), and generally thinly layered

Shear zone A zone where rocks have been deformed and metamorphosed by moving alongside each other

Shield volcano A large volcano, usually formed mainly of basalt, whose diameter is much greater than its height

Silica Silicon dioxide. A fundamental component of all the common rock-forming minerals

Silicic Term referring to a rock or mineral which has a high content of **silica**

Silt Fine-grained sediment whose individual particles range from 0.002 to 0.063 mm in diameter. Silt is coarser than **clay** but finer than sand

Slate A hard, fine grained and finely layered rock formed by the **metamorphism** of **shale**

Spodumene A silicate mineral containing lithium and aluminium

Spreading axis A **plate boundary** where two plates are moving away from each other, and where new material is being emplaced to fill the gap between them. A spreading axis which occurs beneath an ocean is often referred to as a **mid-ocean ridge**

Spring sapping Process by which springs wear back layers of rock between which they are emerging

Surface karst See **karst**

Syenite A coarse-grained igneous rock which has a composition similar to that of granite except that it contains little or no quartz. It is the coarse-grained equivalent of trachyte

Sylvite Potassium chloride

Tantalum A rare metallic element which is important in the manufacture of electronic equipment such as mobile phones, DVD players and computers

Tantalite Oxide of **tantalum**, containing also some iron and manganese

Tectonic Term relating to forces or movements originating within the earth, and applied also to the earth surface features which result from these e.g. **tectonic plates**

Tectonic plate See **plate**

Tillite Material deposited by glaciers, or by meltwater from glaciers, consisting of fragments of mixed rock types and mixed sizes, held in a clayey matrix

Topaz A silicate mineral containing aluminium and fluorine. Topaz is very hard, and if very pure can be used as a gemstone

Tor A small knobby hill made of jointed or broken blocks of rock, frequently **granite**

Tourmaline A complex silicate mineral containing boron. It is usually black, and is sometimes used as a semi-precious gemstone

Trachyte A fine-grained, medium-coloured (often grey) volcanic rock composed mainly of sodium or potassium **feldspar** with a small amount of **quartz** and dark minerals. Its composition is intermediate between those of **basalt** and **rhyolite**

Transform fault A fracture zone between offset segments of a **spreading axis**

Trap Series In Ethiopia, a series of **flood lavas**, largely basalt, deposited over much of the country and parts of the Arabian Peninsula, between about 15 and 45 million years ago

Triple junction A point or region where the boundaries of three tectonic **plates** meet

Tuff General term for finely fragmented material produced in a volcanic eruption. It usually refers to material that is coarser than **ash**

Tuff ring A ring of **tuff** surrounding a central crater. It is formed by a steam explosion, but the source of the eruption, and hence the crater, is less deep than that of a **maar**

Ultramafic Term applied to a rock or mineral which is rick in iron and magnesium and has a low content of **silica**

Unconformity A gap in the rock record. It may represent an absence of deposited material, or **erosion** of material that was deposited

U-shaped valley A valley which has been carved out by a glacier and is U-shaped in cross section, unlike that of a river which has a V-shaped cross section

Vesicle Bubble formed by gas trapped during solidification of **lava**

Vesicular Term describing a rock which contains many **vesicles**

Volcanic rocks See **extrusive rocks**

Watershed A divide that separates one drainage area from another. It is generally a topographic high from which rivers on either side flow in opposite directions

Weathering The process by which rocks are broken down by the action of water, wind, ice or other agencies

Welded tuff Rock formed as the result of a **nuée ardente** eruption. It consists of a mixture of crystals, rock fragments (tuff) and glassy material. It is similar to **ignimbrite** (and the terms are sometimes used synonymously) except that in a welded tuff the fragments are fused together

Zircon Zirconium silicate, a very hard and resistant mineral. Zircons can be used as gems, but in geology they are important as they are resistant to alteration by metamorphism and may record the age and conditions under which they originally crystallised from a magma

Bibliography

This list of references acknowledges the main sources that I have used in preparing this book. Many of them are, however, in specialised journals, out of print or otherwise difficult for the general reader to access. This has been part of my reason for writing this book—to bring together the information which they contain and render it accessible in an easy-to-read form.

General

Asrat A, Demissie M, Mogessie A (2008) Geotourism in Ethiopia. Shama Books, Addis Ababa, 185 pp

Ayenew T (2009) Natural lakes of Ethiopia. Addis Ababa University Press, Addis Ababa, 206 pp

Billi P (ed) (2015) Landscapes and landforms of Ethiopia. Springer, Berlin, 389 pp

Chorowitz J (2005) The East African rift system. J Afr Earth Sci 43:379–410

Corti G (2009) Continental rift evolution: from rift initiation to incipient break-up in the Main Ethiopian Rift, East Africa. Earth Sci Rev 96:1–53

Heldal T, Walle H (2002) Building-stones of Ethiopia. Ethionor mineral resource program (1996–2001) publication (Geological Survey of Ethiopia and Geological Survey of Norway), 61 pp

Mengistu T, Fentaw H (2003) Industrial minerals and rocks resource potential of Ethiopia. Geological Survey of Ethiopia publication, 66 pp

Mohr P (1962) Geology of Ethiopia. University College of Addis, Ababa Press, 268 pp

Mohr P (2009) Africa Beckoning. Millbrook Nova Press, Ireland, 192 pp

Morton W (1978) A field guide to Ethiopian minerals, rocks and fossils. Addis Ababa University Press, Addis Ababa, 170 pp

© Springer International Publishing Switzerland 2016   319
F.M. Williams, *Understanding Ethiopia*, GeoGuide,
DOI 10.1007/978-3-319-02180-5

Tadesse S (2009) Mineral resources potential of Ethiopia. Addis Ababa University Press, Addis Ababa, 290 pp

Maps and Explanatory Notes

CNR and CNRS (ed) (1973) Geological map of the Danakil depression (Northern Afar, Ethiopia) 1:500,000. CNRS (Centre National de la Recherche Scientifique (France)) and CNR (Consiglio Nazionale delle Richerche (Italy)). Map prepared by J. Varet et al
CNR and CNRS (ed) (1975) Geological map of central and southern Afar (Ethiopia and F.T. A.I.) 1:500,000. CNRS (Centre National de la Recherche Scientifique (France)) and CNR (Consiglio Nazionale delle Richerche (Italy))
CNR and CNRS (ed) (1972) Geological map of the Erta Ale volcanic range 1:100,000 CNRS (Centre National de la Recherche Scientifique (France)) and CNR (Consiglio Nazionale delle Richerche (Italy)). Map prepared by F. Barberi and J. Varet
Geological Survey of Ethiopia (1973) Geological map of Ethiopia 1:2,000,000, 1st edn. and explanatory notes
Geological Survey of Ethiopia (1996) Geological map of Ethiopia 1:2,000,000, 2nd edn. and explanatory notes
Geological Survey of Ethiopia (various dates) 1:250,000 geological maps of Ethiopia, complete set with explanatory notes
Lupi L (ed) Afar region Dancalia, geological and route map 1:950,000. Globalmap, Litografia artistica Cartografica, Firenze, Italy
Merla G, Abbate E, Azzaroli A, Bruni P et al (1973) A geological map of Ethiopia and Somalia (1973) 1:2,000,000, with comment and map of major landforms. Firenze, Consiglio Nazionale delle Richerche, Italy

Chapters 1 to 4

For the reader who wishes to delve further into these topics, numerous good introductory textbooks are available, for example "Understanding Earth" by F. Press and R. Siever, 4th ed. 2003, published by W.H. Freeman, 568 pp

Chapter 5

Collins AS, Pisarevsky SA (2005) Amalgamating eastern Gondwana: the evolution of the Circum-Indian Orogens. Earth Sci Rev 71:229–270
Fritz H, Abdelsalam M, Ali KA et al (2013) Orogen styles in the East African Orogen: a review of the Neoproterozoic to Cambrian tectonic evolution. J Afr Earth Sci 86:65–106
Meert J (2003) A synopsis of events related to the assembly of eastern Gondwana. Tectonophysics 362:1–40

Stern RJ (1994) Arc assembly and continental collision in the Neoproterozoic East African Orogen: implications for the consolidation of Gondwanaland. Ann Rev Earth Planet Sci 22:319–351

Chapter 6

Black R, Morton WH, Hailu T (1974) Early structures around the Afar triple junction. Nature 248:496–497
Bussert R (2014) Depositional environments during the Late Palaeozoic ice age in northern Ethiopia, NE Africa. J Afr Earth Sci 99:386–407
Dow DB, Beyth M, Hailu T (1972) Palaeozoic glacial rocks recently discovered in northern Ethiopia. Geol Mag 108:53–60
Gani NDS, Abdelsalam MG, Gera S, Gani MR (2008) Stratigraphic and structural evolution of the Blue Nile Basin, Northwestern Ethiopian Plateau. Geol J 44:30–56
Mitchell RN, Evans AD, Kilian TM (2010) Rapid Early Cambrian rotation of Gondwana. Geology 38:755–758

Chapter 7

Hill R (1991) Starting plumes and continental break-up. Earth Planet Sci Lett 104:398–416
Kumar P, Yuan X, Kumar MR et al (2007) The rapid drift of the Indian tectonic plate. Nature 449:894–897
Ritsema J, Allen RM (2003) The elusive mantle plume. Earth Planet Sci Lett 207:1–12
Witze A (2013) Under the volcano. Nature 504:206–207

Chapter 8

Bosworth W, Huchon P, McClay K (2005) The Red Sea and Gulf of Aden Basins. J Afr Earth Sci 43:334–378
Le Pichon X, Francheteau J (1978) A plate-tectonic analysis of the Red Sea—Gulf of Aden area. Tectonophysics 46:369–406
McKenzie DP, Davies D, Molnar P (1970) Plate tectonics of the Red Sea and East Africa. Nature 226:243–248
Sowerbutts WTC (1972) Rifting in Eastern Africa and the fragmentation of Gondwanaland. Nature 235:435–437
Wolfenden E, Ebinger C, Yirgu G, Renne PR, Kelley SP (2005) Evolution of a volcanic rifted margin: Southern Red Sea, Ethiopia. Geol Soc Am Bull 117:846–864

Chapter 9

Clemens WA, Goodwin MB, Hutchison JH et al (2007) First record of a Jurassic mammal (?"Peramura") from Ethiopia. Acta Palaeontol Pol 52:433–439

Gani et al (2008) (See under Chap. 6)

Goodwin MB, Clemens WA, Hutchison JH et al (1999) Mesozoic continental vertebrates with associated palynostratigraphic dates from the northwestern Ethiopian Plateau. J Vertebr Paleontol 19:728–741

Wolela A (2008) Sedimentation of the Triassic-Jurassic Adigrat Sandstone Formation, Blue Nile (Abay) Basin, Ethiopia. J African Earth Sci 52:30–42

Wolela A (2009) Sedimentation and depositional environments of the Barremian-Cenomanian Debre Libanos Sandstone, Blue Nile (Abay) Basin, Ethiopia. Cretac Res 30:1133–1145

Wolela A (2010) Diagenetic evolution of the Ansian-Pliensbachian Adigrat Sandstone, Blue Nile Basin, Ethiopia. J Afr Earth Sci 56:29–42

Chapter 10

Allen A, Tadesse G (2003) Geological setting and tectonic subdivision of the Neoproterozoic orogenic belt of Tuludimtu, western Ethiopia. J Afr Earth Sci 36:329–343

Alemu T, Hailu K (2013) Field excursion on the Precambrian geology and associated mineralization of Western Ethiopia. In: 24th colloquium of African geology, Addis Ababa, 8–14 Jan 2013

Asrat A, Barbey P, Gleizes G (2001) The Precambrian geology of Ethiopia: a review. Africa Geosci Rev 8:271–288

Blades ML, Collins AS, Foden J et al (In preparation) Age and hafnium isotopic evolution of the Didesa and Kemash domains, western Ethiopia

Woldemichael BW, Kimura J-I, Dunkley DJ et al (2010) SHRIMP U-Pb zircon geochronology and Sr-Nd isotopic systematic of the Neoproterozoic Ghimbo-Nedjo mafic to intermediate intrusions of Western Ethiopia: a record of passive margin magmatism at 855 Ma? Int J Earth Sci 99:1773–1790

Chapter 11

Asrat A (2002) The rock-hewn churches of Tigrai, northern Ethiopia: a geological perspective. Geoarchaeology 17:649–663

Beyth M (1972) Paleozoic-Mesozoic sedimentary basin of Mekele Outlier, northern Ethiopia. Am Assoc Petrol Geol Bull 56:2426–2439

Black et al (1974) (See under Chap. 6)

Ferrari G, Ciampalini R, Billi P, Migon P (2015) Geomorphology of the archaeological area of Aksum (Chap. 7). In: Billi P (ed) (See under General References), pp 147–161

Hagos M, Koeberl C, Kabeto K, Koller F (2010) Geochemical characteristics of the alkaline basalts and the phonolite-trachyte plugs of the Axum area, northern Ethiopia. Austrian J Earth Sci 103:153–170

Natali C, Beccaluva L, Bianchini G, Siena F (2013) The Axum-Adwa basalt-trachyte complex: a late magmatic activity at the periphery of the Afar plume. Contrib Miner Petrol 166:351–370

Nyssen J, Frankl A, Munro RN et al (2010) Digital photographic archives for environmental and historical studies: an example from Ethiopia. Scott Geogr J 126:185–207

Chapter 12

Abbate E, Bruni P, Ferretti MP et al (2014) The East African Oligocene intertrappean beds: regional distribution, depositional environments and Afro/Arabian mammal dispersals. J Afr Earth Sci 99:463–489

Abbate E, Bruni P, Sagri M (2015) Geology of Ethiopia: a review and geomorphological perspectives (Chap. 2). In: Billi P (ed) (See under General References), pp 33–64

Abebe T, Mazzarini F, Innocenti F, Manetti F (1998) The Yerer-Tullu Wellel volcanotectonic lineament: a trans-tensional structure in central Ethiopia and the associated magmatic activity. J Afr Earth Sci 26:135–150

Beccaluva L, Bianchini G, Natali C, Diena F (2009) Continental flood basalts and mantle plumes: a case study of the northern Ethiopian plateau. J Petrol 50:1377–1403

Bonnefille R (2010) Cenozoic vegetation, climate changes and hominid evolution in tropical Africa. Glob Planet Change 72:390–411

Hofmann C, Courtillot V, Féraud G et al (1997) Timing of the Ethiopian flood basalt event and implications for plume birth and global change. Nature 389:838–841

Hurni H (1982) Climate and the dynamics of altitudinal belts from the last cold period to the present day. In: Semien Mountains, Ethiopia. Geographica Bernensia G13, vol II. Beheft 7 zum Jahrbuch der Geographischen Gesellschaft von Bern, 197 pp, plus folder of maps

Kappelman J, Rasmussen DT, Sanders WJ et al (2003) Oligocene mammals from Ethiopia and faunal exchange between Afro-Arabia and Eurasia. Nature 426:549–552

Kieffer B, Arndt N, Lapierre H et al (2004) Flood and shield basalts from Ethiopia: magmas from the African superswell. J Petrol 45:793–834

Mazzero F, Desagulier C, Rondeau B et al (2010) L'opale du Wollo, Ethiopie: des mines de gisement! Revue de l'Association Francaise de Gemnologie 174:14–20

Mège D, Korme T (2004) Dyke swarm emplacement in the Ethiopian Large Igneous Province: not only a matter of stress. J Volcanol Geoth Res 132:283–310

Mohr P (1983) Ethiopian flood basalt province. Nature 303:577–584

Pik R, Deniel C, Coulon C et al (1998) The northwestern Ethiopian Plateau flood basalts: classification and spatial distribution of magma types. J Volcanol Geoth Res 81:91–111

Sanders WJ, Kappelman J, Rasmussen DT (2004) New large-bodied mammals from the late Oligocene site of Chilga, Ethiopia. Acta Palaeontol Pol 49:365–392

Umer M, Kebede S, Osmaston H (2004) Quaternary glacial activity on the Ethiopian mountains. In: Ehlers J, Gibbard PL (eds) Quaternary glaciations—extent and chronology, part III. Elsevier, Amsterdam, pp 171–174

Chapter 13

Gani NDS, Gani MR, Abdelsalam MG (2007) Blue Nile incision on the Ethiopian Plateau: pulsed plateau growth, Pliocene uplift, and hominin evolution. GSA Today 17:4–11

Gani NDS, Abdelsalam MG (2006) Remote sensing analysis of the Gorge of the Nile, Ethiopia, with emphasis on Dejen-Gohatsion region. J Afr Earth Sci 44:135–150

Cheesman RE (1935) Lake Tana and its islands. Geogr J 85:489–502

Chorowitz J, Collet B, Bonavia FF et al (1998) The Tana basin, Ethiopia: intra-plateau uplift, rifting and subsidence. Tectonophysics 295:351–367

Grabham GW, Black RP (1925) Report of the mission to Lake Tana 1920–1921. Government Press, Cairo

Hautot S, Whaler K, Gebru W, Desissa M (2006) The structure of a Mesozoic basin beneath the Lake Tana area, Ethiopia, revealed by magnetotelluric imaging. J Afr Earth Sci 44:331–338

Kebede S (2013) Groundwater in Ethiopia: features, numbers and opportunities (Chap. 2). In: Groundwater occurrence in regions and basins. Springer, Berlin, pp 15–121 (287 pp)

Lamb HF, Bates CR, Coombes PV et al (2007) Late Pleistocene desiccation of Lake Tana, source of the Blue Nile. Quat Sci Rev 26:287–299

Lamb HF, Bates CR, Bryant CL et al (In preparation) Climatic context of modern human dispersal from north-east Africa

Marshall MH, Lamb HF, Huws D et al (2011) Late Pleistocene and Holocene drought events at Lake Tana, the source of the Blue Nile. Global Planet Change 78:147–161

Oestigaard T, Gedel AF (2011) Gish Abay: the source of the Blue Nile. WIT Trans Ecol the Environ 153:27–38

Pik R, Marty B, Carignan J, Lavé J (2003) Stability of the Upper Nile drainage network (Ethiopia) deduced from (U-Th)/He thermochronometry: implications for uplift and erosion of the Afar plume dome. Earth Planet Sci Lett 215:73–88

Poppe L, Frankl A, Poesen J et al (2013) Geomorphology of the Lake Tana basin, Ethiopia. J Maps 9(3):431–437

Chapter 14

Asrat et al (2008) (See General References)

Asrat A, Ayallew Y (2011) Geological and geotechnical properties of the medieval rock hewn churches of Lalibela, northern Ethiopia. J Afr Earth Sci 59:61–73

Chapter 15

Asrat, A (2015) Geology, geomorphology, geodiversity and geoconservation of the Sof Omar Cave System, southeastern Ethiopia. J Afr Earth Sci 108:47–63

Assefa Z, Pleurdeau D, Duquesnoy F et al (2014) Survey and explorations of caves in southeastern Ethiopia: Middle Stone Age and Later Stone Age archaeology and Holocene rock art. Quat Int 343:136–147

Asrat A, Baker A, Umer M et al (2007) A high-resolution multi-proxy stalagmite record from Mechara, Southeastern Ethiopia: palaeohydrological implications for speleothem palaeoclimate reconstruction. J Quat Sci 22:53–63

Black et al (1974) (See under Chap. 6)

Bosellini A, Russo A, Assefa G (2001) The Mesozoic succession of Dire Dawa, Harar Province, Ethiopia. J Afr Earth Sci 32:403–417

Catlin D, Largen MJ, Monod T, Morton WH (1973) The caves of Ethiopia. Trans Cave Res Group G B 15:108–168

Clark JD, Harris JWK (1985) Fire and its roles in early hominid lifeways. Afr Archaeol Rev 3:3–27

Hunegnaw A, Sage L, Gonnard R (1998) Hydrocarbon potential of the intracratonic Ogaden Basin, SE Ethiopia. J Petrol Geol 21:401–425

Mège D, Purcell P, Bezos A et al (2015) A major dike swarm in the Ogaden region south of Afar and the early evolution of the Afar Triple Junction. In: Wright TJ, Ayele A, Ferguson D, Kidane T, Vye-Brown C (eds) Magmatic rifting and active volcanism. Geological Society, London, Special Publication no. 420

Mège D, Purcell PG, Pochat S, Guidat T (2015) The landscape and landforms of the Ogaden, southeast Ethiopia (Chap. 19). In: Billi P (ed) (See under General References), pp 323–348

Osmaston HA, Mitchell WA, Osmaston JAN (2005) Quaternary glaciation of the Bale Mountains, Ethiopia. J Quat Sci 20:593–606

Potter EC (1976) Pleistocene glaciation in Ethiopia: new evidence. J Glaciol 17:148–150

Purcell PG (1976) The Marda Fault Zone, Ethiopia. Nature 261:569–571

Purcell PG (1979) The geology and petroleum potential of the Ogaden Basin, Ethiopia. Unpublished report, 76 pp

Umer M, Kebede S, Osmaston H (2004) (See under Chap. 12)

Williams MAJ, Williams FM, Gasse F et al (1979) Plio-Pleistocene environments at Gadeb prehistoric site, Ethiopia. Nature 282:29–33

Worku T, Astin TR (1992) The Karoo sediments (Late Palaeozoic to Early Jurassic) of the Ogaden Basin, Ethiopia. Sediment Geol 76:7–21

CHAPTER 16

Abebe B, Acocela V, Korme T, Ayalew D (2007) Quaternary faulting and volcanism in the Main Ethiopian rift. J Afr Earth Sci 48:115–124

Accademia Nazionale dei Lindei (1980) Geodynamic evolution of the Afro-Arabian Rift System. In: Proceedings of international meeting organised by the Accademia Nazionale dei Lincei and the Consiglio Nazionale delle Richerche. Rome, 18–20 Apr 1979, 705 pp

Acocella V, Korme T, Salvini F (2003) Formation of normal faults along the axial zone of the Ethiopian Rift. J Struct Geol 25:503–513

Baker BH, Mohr PA, Williams LAJ (1972) Geology of the Eastern Rift System of Africa. Special Report Geological Society of America, vol 136, 67 pp

Bilham R, Bendick R, Larson K et al (1999) Secular and tidal strain across the Main Ethiopian Rift. Geophys Res Lett 26:2789–2792

Corti G (2009) (See under General References)

Ebinger CJ, Casey M (2001) Continental break-up in magmatic provinces: an Ethiopian example. Geology 29:527–530

Korme T, Acocella V, Abebe B (2004) The role of pre-existing structures in the origin, propagation and architecture of faults in the Main Ethiopian Rift. Gondwana Res 7:467–479

Mohr P (1987) Patterns of faulting in the Ethiopian rift valley. Tectonophysics 143:169–179

Mohr P (1967) The Ethiopian rift system. Bull Geophys Obs Addis Ababa 11:1–65

Pan M, Sjöberg LE, Asfaw LM et al (2002) An analysis of the Ethiopian rift valley GPS campaigns in 1994 and 1999. J Geodyn 33:333–343

UNDP (1973) Geology, geochemistry and hydrology of hot springs of the East African Rift System within Ethiopia. Technical report prepared for the Imperial Ethiopian Government for the United Nations Development Programme, United Nations, New York, 227 pp + maps

Chapter 17

Benvenuti M, Carnicelli S, Belluomini G et al (2002) The Ziway-Shala lake basin (main Ethiopian rift, Ethiopia): a revision of basin evolution with special reference to the late quaternary. J Afr Earth Sci 35:247–269

Boccaletti M, Bonini B, Mazzuoli R, Trua T (1999) Pliocene-Quaternary volcanism and faulting in the northern Main Ethiopian Rift (with two geological maps at scale 1:50,000). Acta Vulcanol 11:83–97

Boccaletti M, Mazzuoli R, Bonini M et al (1999) Plio-Quaternary volcanotectonic activity in the northern sector of the Main Ethiopian Rift: relationships with oblique rifting. J Afr Earth Sci 29:679–698

Corti et al (2013) (See under Chap. 20)

Di Paola GM (1972) The Ethiopian Rift Valley (between 7°00' and 8°40' lat. north). Bull Volcanol 36:517–560

Di Paola GM (1971) Geology of the Corbetti Caldera area (Main Ethiopian Rift Valley). Bull Volcanol 35:497–506

Le Turdu C, Tiercelin J-J, Gibert E et al (1999) The Ziway-Shala lake basin system, Main Ethiopian Rift: Influence of volcanism, tectonics, and climatic forcing on basin formation and sedimentation. Palaeogeogr Palaeoclimatol Palaeoecol 150:135–177

Levitte D, Columba J, Mohr P (1974) Reconnaissance geology of the Amaro Horst, Southern Ethiopian Rift. Geol Soc Am Bull 85:417–422

Macdonald R, Gibson IL (1969) Pantelleritic obsidians from the volcano Chabbi (Ethiopia). Contrib Mineral Petrol 24:239–244

Ministry of Water Resources (2008) Butajira-Ziway areas development study. Federal Democratic Republic of Ethiopia, Ministry of Water Resources, Ethiopian Water Technology Centre Report, Jan 2008, 97 pp

Mohr P (1959) Report on a geological excursion through Southern Ethiopia. Bull Geophys Obs Addis Ababa 2:9–19

Mohr PA (1966) Chabbi volcano (Ethiopia). Bull Volcanol 29:797–815

Mohr P, Mitchell JG, Raynolds RGH (1980) Quaternary volcanism and faulting at O'a caldera, central Ethiopian Rift. Bull Volcanol 43:173–189

Sagri M, Bartolini C, Billi P et al (2008) Latest Pleistocene and Holocene river network evolution in the Ethiopian Lakes Region. Geomorphology 94:79–97

Woldegabriel G, Aronson JL, Walter RC (1990) Geology, geochronology, and rift basin development in the central sector of the Main Ethiopia Rift. Geol Soc Am Bull 102:439–458

Woldegabriel G, Yemane T, Suwa G et al (1991) Age of volcanism and rifting in the Burji-Soyama area, Amaro Horst, southern Main Ethiopian Rift: geo-and biochronologic data. J Afr Earth Sci 13:437–447

Chapter 18

Asfaw B, Beyene Y, Suwa G et al (1992) The earliest Acheulean from Konso-Gardula. Nature 360:732–735

Atnafu B, Bonavia FF (1991) Precambrian structure and Late Pleistocene strike-slip tectonics around Mega (southern Ethiopia). J Afr Earth Sci 13:527–530

Levitte D, Columba J, Mohr P (1974) Reconnaissance geology of the Amaro Horst. Geol Soc Am Bull 85:417–422

Philippon M, Corti G, Sani F et al (2014) Evolution, distribution and characteristics of rifting in southern Ethiopia. Tectonics 33:485–508

Shinji N, Katoh S, Woldegabriel G et al (2005) Lithostratigraphy and sedimentary environments of the hominid-bearing Pliocene-Pleistocene Konso Formation in the southern Main Ethiopian Rift, Ethiopia. Palaeogeogr Palaeoclimatol Palaeoecol 216:333–357

Suwa G, Nakaya H, Asfaw B et al (2003) Plio-Pleistocene terrestrial mammal assemblage from Konso, southern Ethiopia. J Vertebr Paleontol 23:901–916

Tadesse S (2009) (See under General References)

Woldegebriel G, Yemane T, Suwa G et al (1991) Age of volcanism and rifting in the Burji-Soyoma area, Amaro Horst, southern Main Ethiopian Rift: geo- and biochronologic data. J Afr Earth Sci 13:437–447

Chapter 19

Cole JW (1969) Gariboldi volcanic complex, Ethiopia. Bull Volcanol 33:566–578

Gibson IL (1967) Preliminary account of the volcanic geology of Fantale, Shoa. Bull Geophys Obs Addis Ababa 10:59–67

Gibson IL, Tazieff H (1967) Additional theory of origin of fiamme in ignimbrites. Nature 215:1473–1474

Gibson IL (1969) The structure and volcanic geology of an axial portion of the Main Ethiopian Rift. Tectonophysics 8:561–565

Gibson IL (1970) A pantelleritic welded ash-flow tuff from the Ethiopian Rift Valley. Contrib Mineral Petrol 28:89–111

Gibson IL (1974) Blister caves associated with an Ethiopian volcanic ash-flow tuff. Stud Speleol 2:225–237

Goerner A, Jolie E, Gloaguen R (2009) Non-climatic growth of the saline Lake Besaka, Main Ethiopian Rift. J Arid Environ 73:287–295

Harris WC (1844) The highlands of Ethiopia, vol 3 (Chaps. 29 and 30). Longman, Brown and Longman. London

Kebede S (unpublished) Origin of water and geochemical mixing model: geochemical and isotopic evidence. Lake Besaka interim report. Addis Ababa University

Kidane T, Otofuji Yo-I, Komatsu Y, Shibasaki H (2009) Paleomagnetism of the Fentale-magmatic segment, main Ethiopian Rift: New evidence for counterclockwise block rotation linked to transtensional deformation. Phys Earth Planet Inter 176:109–123

McBirney A (1968) Second additional theory of origin of fiamme in ignimbrites. Nature 217:938

Rampey ML, Oppenheimer C, Pyle DM, Yirgu G (2010) Caldera-forming eruptions of the Quaternary Kone Volcanic Complex, Ethiopia. J Afr Earth Sci 58:51–66

Tazieff H, Gibson I (1967) Sur la genèse de l'ignimbrite pantelléritique de Fantalé. Comptes Rendues de l'Académie des Sciences, Paris 265:950–953

Williams FM, Williams MAJ, Aumento F (2004) Tensional fissures and crustal extension rates in the northern part of the Main Ethiopian Rift. J Afr Earth Sci 38:183–197

Wolfenden E, Ebinger C, Yirgu G et al (2004) Evolution of the northern Main Ethiopian Rift: birth of a triple junction. Earth Planet Sci Lett 224:213–228

Chapter 20

Corti G, Sani F, Phillipon M et al (2013) Quaternary volcano-tectonic activity in the Soddo region, western margin of the Southern Main Ethiopian Rift. Tectonics 32:861–879

www.ancient-origins.net/africa. The intricately carved Tiya megaliths of Ethiopia

Zanettin B, Justin-Visentin E (1974) The volcanic succession in central Ethiopia. The volcanics of the western Afar and Ethiopian rift margins. Memoir of the Institute of Geology and Mineralogy of the University of Padua, vol 31, 20 pp

Chapter 21

Barberi F, Tazieff H, Varet J (1972) Volcanism in the Afar depression: its tectonic and magmatic significance. Tectonophysics 15:19–29

Barberi F, Borsi S, Ferrara G et al (1972) Evolution of the Danakil Depression (Afar, Ethiopia) in light of radiometric age determinations. J Geol 80:720–729

Berhe S (1986) Geologic and geochronologic constraints on the evolution of the Red Sea-Gulf of Aden and Afar Depression. Journal of African Earth Sciences 5:101–117

Beyene A, Abdelsalam MG (2005) Tectonics of the Afar depression: a review and synthesis. J Afr Earth Sci 41:41–59

Lupi L (2008) Dancalia: L'ezplorazione dell'Afar, un'avventura italiana, vol 1. Instituto Geografico Militare, Firenze, 687 pp

Lupi L (2009) Dancalia: L'ezplorazione dell'Afar, un'avventura italiana, vol 2. Instituto Geografico Militare, Firenze, 1487 pp

Makris J, Ginzburg A (1987) The Afar Depression: transition between continental rifting and sea-floor spreading. Tectonophysics 41:199–214

Mohr P (1989) Nature of the crust under Afar: new igneous, not thinned continental. Tectonophysics 167:1–11

Yirgu G, Ebinger CJ, Maguire PKH (eds) (2006) The Afar volcanic province within the East African Rift System. Geological Society Special Publication 259. The Geological Society, London, 327 pp

Pilger A, Rösler A (eds) (1975) Afar depression of Ethiopia. In: Proceedings of an international symposium on the Afar region and related rift problems held in Bad Bergzabern F.R. Germany, Apr 1–6, 1974. Inter-Union Commission on geodynamics scientific report no. 14. E. Schweizerbart'sche Verlagsbuchhandlung (Nägele u. Obermiller) Stuttgart, 416 pp

Pilger A, Rösler A (eds) (1976) Afar between Continental and Oceanic Rifting. In: Proceedings of an international symposium on the Afar region and related rift problems held in Bad Bergzabern F.R. Germany, April 1–6, 1974. Inter-Union Commission on Geodynamics Scientific Report No. 16. E. Schweizerbart'sche Verlagsbuchhandlung (Nägele u. Obermiller) Stuttgart, 216 pp

Wolfenden E, Ebinger C, Yirgu G et al (2005) Evolution of a volcanic rifted margin: southern Red Sea, Ethiopia. Geol Soc Am Bull 117:846–864

Chapter 22

Acocella V (2006) Regional and local tectonics at Erta Ale caldera. J Struct Geol 28:1808–1820

Ayele A, Jacques E, Kassim M (2007) The volcano-seismic crisis in Afar, Ethiopia, starting Septmeber 2005. Earth Planet Sci Lett 255:177–187

Barberi F, Varet J (1970) The Erta Ale Volcanic Range (Danakil Depression, Northern Afar, Ethiopia). Bull Volcanol 34:848–917

Bonatti E, Tazieff H (1970) Exposed guyot from the Afar Rift, Ethiopia. Science 168:1087–1089

Bonatti E, Emiliani C, Ostlund G, Rydell H (1971) Final desiccation of the Afar Rift, Ethiopia. Science 172:468–469

CNR-CNRS Afar Team (1973) Geology of northern Afar, Ethiopia. Revue de Géographie Physique et de Géologie Dynamique (Paris) 15:443–490

Ebinger C (2006) The birth of an ocean. Planet Earth Autumn 2006:26–27

Ferguson DJ, Barnie TD, Pyle DM et al (2010) Recent rift-related volcanism in Afar, Ethiopia. Earth Planet Sci Lett 292:409–418

Field L, Blundy J, Calvert A, Yirgu G (2013) Magmatic history of Dabbahu, a composite volcano in the Afar Rift, Ethiopia. Geol Soc Am Bull 125:128–147

Franzson H, Helgadóttir HM, Óskarsson F (2015) Surface exploration and first conceptual model of the Dallol Geothermal Area, Northern Afar, Ethiopia. In: Proceedings of the World Geothermal Congress, Melbourne, Australia, 19–25 Apr 2015, 11 pp

Hamling I, Wright TJ, Calais E et al (2010) Stress transfer between thirteen successive dyke intrusions in Ethiopia. Nature Geosci 3:713–717

Holwerda JG, Hutchinson RW (1968) Potash-bearing evaporites in the Danakil area, Ethiopia. Econ Geol 63:124–150

Keir D (2008) Watching from Afar. Planet Earth, Autumn 2008:16–17

Keir D, Bastow ID, Pagli C, Chambers EL (2013) The development of extension and magmatism in the Red Sea rift of Afar. Tectonophysics 607:98–114

Oppenheimer C, Francis P (1998) Implications of longeval lava lakes for geomorphological and plutonic processes at Erta Ale volcano, Afar. J Volcanol Geoth Res 80:101–111

Pagli C, Wright TJ, Ebinger CJ et al (2012) Shallow axial magma chamber at the slow-spreading Erta Ale Ridge. Nature Geosci 5:284–288

Sigmundsson F (2006) Magma does the splits. Nature 442:251–252

Tazieff H (1973) The Erta Ale volcano. Revue de Géographie Physique et de Géologie Dynamique 15:437–441

United Nations Development Programme (1973) Geology, geochemistry and hydrology of hot springs of the East African Rift system within Ethiopia. Technical report, United Nations, New York

Wright T (2006) Making the paper. Nature 442:xi

Wright T, Sigmundsson F, Pagli C et al (2012) Geophysical constraints on the dynamics of spreading centres from rifting episodes on land. Nature Geosci 5:242–250

Chapter 23

Acocella V, Abebe B, Korme T, Barberi F (2008) Structure of Tendaho Graben and Manda Hararo Rift: implications for the evolution of the southern Red Sea propagator in Central Afar. Tectonics 27:TC4016. doi:10.1029/2007TC002236

Bridges DL, Mickus K, Gao SS et al (2012) Magnetic stripes of a transitional continental rift in Afar. Geol Soc Am Bull 40:203–206

Civetta L, de Fino M, Gasparini P et al (1975) Structural meaning of east-central Afar volcanism (Ethiopia, T.F.A.I.). J Geol 83:363–373

De Fino M, La Volpe L, Lirer L, Varet J (1973) Geology and petrology of Manda-Inakir Range and Moussa Alli volcano, central eastern Afar (Ethiopa and T.F.A.I.). Revue de Géographie Physique et de Géologie Dynamique 15:373–386

Gibson IL, Tazieff H, Hepworth JV (1970) The structure of Afar and the northern part of the Ethiopian Rift (and discussion). Philos Trans R Soc Lond Ser A Math Phys Sci 267:331–338

Lemma Y, Thiel S, Heinson G (2015) Three dimensional conductivity model of the Tendaho high enthalpy geothermal field. J Volcanol Geoth Res 290:53–62

Lemma Y, Kalberkamp U, Abera F et al (2012) Magnetotelluric exploration at Tendaho High Temperature Geothermal Field in North East Ethiopia. GRC Trans 36:9–12

Tesfaye S, Harding DJ, Kusky T (2003) Early continental breakup boundary and migration of the Afar triple junction, Ethiopia. Geol Soc Am Bull 115:1053–1067

Varet J, Gasse F (1978) Geology of central and southern Afar (Ethiopia and Djibouti Republic). Editions du Centre National de la Recherche Scientifique, 124 pp

Vellutini P (1990) The Manda-Inakir Rift, Republic of Djiboui: a comparison with the Asal Rift and its geodynamic interpretation. Tectonophysics 172:141–153

Chapter 23 (Palaeo-anthropology)

Haile-Sellassie Y, Saylor BZ, Deino A et al (2012) A new hominin foot from Ethiopia shows Pliocene bipedal adaptations. Nature 483:565–569

Johanson DC, Edey MA (1981) Lucy, the beginnings of humankind. Granada Publishing, London, 409 pp

Johanson DC, Wong K (2010) Lucy's Legacy: the quest for human origins. Three River Press, New York, 321 pp

Kalb J (2000) Adventures in the bone trade. Springer, New York, 389 pp

Lieberman DE (2012) Those feet in ancient times. Nature 483:550–551

Semaw S, Renne P, Harris JWK (1997) 2.5 million-year-old stone tools from Gona, Ethiopia. Nature 385:333–336

Taieb M, Johanson DC, Coppens Y, Aronson JL (1976) Geological and palaeontological background of Hadar hominid site, Afar, Ethiopia. Nature 260:289–293

Taieb M (1985) Sur la terre des premiers hommes. Robert Laffont, Paris, 330 pp

Walter RC (1994) Age of Lucy and the First Family: single-crystal ^{40}Ar/^{39}/Ar dating of the Denen Dora and lower Kada Hadar members of the Hadar Formation, Ethiopia. Geology 22:6–10

White TD (1986) Cut marks on the Bodo cranium: a case of prehistoric defleshing. Am J Phys Anthropol 69:503–509

White TD, Suwa G, Asfaw B (1994) Australopithecus ramidus, a new species of early hominid from Aramis, Ethiopia. Nature 371:306–312

Chapter 24

Dakin F, Gouin P, Searle R (1971) The 1969 earthquakes in Serdo (Ethiopia). Bull Geophys Obs Addis Ababa 13:19–56

Gouin P (1975) Karakore and Serdo epicentres: relocation and tectonic implications. Bull Geophys Obs Addis Ababa 15:15–25

Gouin P (1969) Earthquake history of Ethiopia and the Horn of Africa. International Development Research Centre, Ottawa, 258 pp

Kebede F, Kim W-Y, Kulhánek O (1989) Dynamic source parameters of the March-May 1969 Serdo earthquake sequence in central Afar, Ethiopia, deduced from teleseismic body waves. J Geophys Res 94:5603–5614

Chapter 25

Bosworth et al (2005) (See under Chap. 8)

Buiter SJH, Torsvik TH (2014) A review of Wilson Cycle plate margins: a role for mantle plumes in continental break-up along sutures. Gondwana Res 26:627–653

Index of Localities

A

Abaya, Lake, 191, 193, 196
Abay, River. *See* Blue Nile
Abhe, Lake, 273
Abiata, Lake, 187–189
Abuna Yemata Guh, church (Lalibela), 89–91
Adado Graben, 274, 276
Adama. *See* Nazret
Addis Ababa, 118–119, 229–230, 290–291
Aden, Gulf of, 47, 49, 174, 249–250, 278, 296–298, 301
Adi Arkay, 70, 84
Adigrat, 36, 84–85
Adola. *See* Kibre Mengist
Adua, 95–98
Adua, Battle of, 98
Afar, 21, 47, 51, **243–247**, **249–270**, **271–283**, 285–289, 296, 297
Afar, central and southern, 271–283
Afar, northern, 249–270
Afar, various names for, 243
Afar Window, 237
Afara Dara, Mt, 269
Afdem, Mt, 235
Afdera, Mt, 265, 269
Afrera, Lake, 258, 264
Ahmar, Mts, 155, 156
Alayta, Mt, 265
Alayta, volcanic range, 251, 264, 265
Ale Bagu, Mt, 253
Alid, Mt, 262

Alu, Mt, 253
Alutu, Mt, 185
Amaro Horst. *See* Amaro Mts
Amaro, Mts, 193, **195–196**
Amba Alaji, Mt, 239, 241
Amba Aradam, 87
Ambo, 118–119
Ankober, 289–290
Aqaba, Gulf of, 172
Ara Shatan, Lake, 234
Arabian Peninsula, 41, 43, 47, 247, 298
Aramis, 282
Aranguadi, Lake (Debre Zeit), 231, 232
Asa Ale, Mt, 269
Asal, Lake (Djibouti), 245
Asal rift, 275, **278**
Ashengi, Lake, 239, 240
Asheton Mariam, church (Lalibela), 143, 145
Asheton, Mt, 143
Asosa, 82
Asseb, 290
Assebot, Mt, 235
Atbara, River, 67
Awasa, Lake, 187, **190**
Awash, 209, 211
Awash Falls, 218, 221
Awash gorge, 218, 222
Awash National Park, 213–219
Awash, River, 221, 232, 277
Axum, 98–101
Ayelu, Mt, 276

© Springer International Publishing Switzerland 2016
F.M. Williams, *Understanding Ethiopia*, GeoGuide,
DOI 10.1007/978-3-319-02180-5

B
Bab el Mandeb, Strait of, 47, 249, 279
Babu Gaya, Lake, 231
Badda, Mt, 157, 159
Bahir Dar, 126
Bakili, Lake, 260
Bale Mts, 157–161
Batu, Mt, 159
Belekiya, Mts, 240–242
Beles, River, 129
Berhale, 240
Bericha, Mt, 185
Besaka, Lake, 213, **219–220**, **222**, 223
Bidu, volcanic complex, 269
Bilbala, 145
Bishoftu. *See* Debre Zeit
Bishoftu, Lake, 231
Blue Nile, Gorge, **57–67**, 131–132
Blue Nile, River, 57, **130–133**
Blue Nile, Source of, 128–130
Bodo (Palaeo-anthropolocical site), 282
Boina. *See* Dabbahu
Borale Ale, Mt, 253
Borawli, Mt., 252
Borkenna Graben, 239, 290
Boseti Bericha, Mt, 210
Boseti Guda, Mt, 210
Bridge of God. *See* Tosa Sucha
Buia Graben, 240
Bulbula, River, 190, 191
Bure, 124
Butajira, 180, 228, 234, 291
Buyit, 115

C
Chabbi, Mt, 185, 186, 190
Chamo, Lake, 187, 191, 193, 196
Checheho, 147, 151
Cheleleka, Lake, 231
Chencha Mts, 234
Chenek (Semien Mts.), 115, 116
Chercher, Mts, 155
Chew Bahir, Lake, 198, 203
Chilalo, Mt, 157, 159, 227, 230
Chilga, 107, 108

Choke, Mt, 111, 122, 124, 130
Corbetti caldera, 190

D
Dabbahu (Boina), Mt, **267**, 292
Dakhata Valley. *See* Valley of Marvels
Dalafilla, Mt, 253
Daliti, 78
Dallol, **262–264**, 303
Dama Ale, Mt, 278
Damot. *See* Yeha
Damota, Mt, 234
Danakil. *See* under Afar
Danakil Alps, 86, **268**, 301
Da'Ure, Mt, 267
Dawa, River, 169, 206
Dead Sea, 245
Dead Sea Rift, 172, 173
Debarek, 68, 70
Debre Damo, 95, 96
Debre Libanos, 65, 68
Debre Sina, 107, 237, 238
Debre Zeit (Bishoftu), 230–232
Dek, Island (Lake Tana), 127, 128
Dendi, Mt, 119
Dessa, cave, 163
Dessie, 239
Didessa, River, 79, 129
Dikika, 282
Dire Dawa, 157, 235
Djibouti, 245, 249
Dobi Graben, 273, 275, 276

E
Edaga Arbi, 85
El Sod, crater, **203–204**, 205
Encuolo, Mount, 157, 159, 227
Enda Arbi Guna, 70
Enda Selassie (Shire), 85
Enticho, 85, 95
Entoto, Mt, 118, 229
Eritrea, 19, 85, 237, 238, 245, 269
Erta Ale, Mt, 251, **253–258**
Erta Ale, volcanic range, **251–253**, 262, 264

F

Fafan, River, 169
Fantale, Mt, 210, **213–218**, 223, 303
Filweha (Addis Ababa), 119
Filweha (Awash National Park), 218, 220
Furi, Mt, 230, 234

G

Gabilerna, Mt, 244, 274
Gada Ale, Mt, 252, 253
Gadeb, 161
Gademota, caldera, 187
Galana Graben, 196
Gamarri scarp, 273
Gambela, 81
Ganale, River, 161, 169
Ganeta Mariam, church (Lalibela), 145
Ganjuli Graben, 196
Gara Adi, Mt, 235
Garba Guracha, lake (Bale Mts.), 158
Garsat Graben, 240, 241
Gedamsa, Caldera, 193
Gestro, River, 161, 169
Gewani, 276
Gheralta, Mt. (Tigray), 89, 92, 93
Ghoubbet el Kharab, 278
Gimbi, 77, 78, 80
Gish Abay, 128–130
Gobad Graben, 276
Gobedra, hill, 100, 101
Goha Tsion, 59–61, 63
Gona (Palaeo-anthropological site), 282
Gonder, 124
Guguftu, Mt, 111
Gumbi, Mt, 235
Guna, Mt, 111, 112, 122, 124, 131
Gurage, Mts, 180, 232, 234

H

Hadar, 279, 282
Harenna, Forest, 161
Harer, 164, 165
Hauzien, district, 85, 87–94
Hayk Graben, 239
Hayk, Lake, 239

Hayli Gub, Mt, 253
Herto-Bouri (Palaeo-anthropological site), 282
Hintalo (Antalo), 38, 61, 86
Hora, Lake, 231
Horo Kelo, River, 189

I

Ilbah (Ogaden region), 168
Imrahana Christos, church (Lalibela), 146, 149, 150
India, 11, 29, 30, 34, 39, 41, 45, 49, 298
Injibara, 125

J

Jamma, River, 60, 65, 68
Jijiga, 165, 166
Jima, 117

K

Kaka, Mt, 157, 159, 227, 228
Karakore, 290–291
Karum, Lake, 260, 262
Kenticha Mine, 207, 208
Kibre Mengist, 205, 207
Kilole, Lake, 231
Kobo Graben, 239
Koftu, Lake, 231
Koka Dam, 219
Koka, Lake, 187
K'one, Caldera, 210, **211–213**, 223
Konso, 200
Koriftu, Lake, 231
Korkor Mariam, church. *See* Mariam Korkor
Kurmuk, 81, 82
Kurub, Mt, 278

L

Lalibela, 67, **135–151**
Langano, Lake, 183, **187–189**, 291
Ledi-Geraru (Palaeo-anthropoligical site), 282

Lega Dembi Mine, 206
Lima Limo escarpment, 68
Limmo (granite intrusion), 269
Little Abay, River, 128–130
Logia, 277, 278

M
Ma Alalta, Mt, 269
Mai Shaha, River, 113, 115
Maichew, 239, 240
Manda Hararo, volcanic range, 251,
    265–268, 275, 277, 288
Manda Inakir, volcanic range, 278
Marda Pass, 166
Marda Range, **165–166**, 167
Mariam Korkor, church, 93, 94
Mechara, 162, 164
Mediterranean Sea, 47, 303
Mega, 198, 201
Mega Graben, 205
Mega Horst, 205
Megezez, Mt, 237
Mekdela, 104
Mekele, 85
Mekele, Basin, 84, **85–87**
Melka Kunture, 232–233
Mangestu, Mt, 22, 111, 112, 122, 131
Merowe, 103
Mikael Kurara, church, 91, 94
Mille, 272, 277
Millennium Bridge, 59, 60, 66, 131
Mojo, River, 192
Moussa Ale, Mt, 279
Mugher, River, 65

N
Nabro, Mt, 269
Nakuta La'ab, church, 138, 143, 147, 148
Narga Sellassie, Monastery, 128
Nazret, 209, 211
Nech Sar National Park, 191, 194
Nejo, 74
Nekemte, 73
"New York", 200–202
Nile Valley, 132

O
Ogaden, 155, **167–170**
Omo Rift, 175
Omo, River, 197

R
Ras Dashen, Mt, 112, 115, 159
Red Sea, 38, 47, 49, 166, 172, 174, 177,
    247, 249, 251, 259, 270, 293,
    296–298, 301–302
Regghae Badda, volcanic complex, 119
Robit Graben, 238, 239
Roha, 135

S
Sabober, tuff ring, 217, 219
Salt Plain, 258–262
Sankober, 112, 115
Semera, 268, 288
Semien, Mts, 44, 67–70, **112–117**
Serdo, 278, **285–289**
Shakiso, 206, 207
Shala, Lake, 183, 187–190
Shashamane, 185
Shebeli, River, 155, 161, 162, 169, 170
Sheik Hussein, 162
Shire. See Enda Sellassie
Sire, 227, 234
Sodo, 228, 234
Sof Omar, cave, 161–163
Southeastern Highlands, 19, 20, **153–157**
Suez, Gulf of, 47

T
Tadjura, Gulf of, 245, 249, 278
Tana, Lake, 57, **121–127**, 129
Tat Ale volcanic range, 251, 264, 265
Tekeze, River and Gorge, **67–71**, 83,
    113–115
Tendaho Dam, 277
Tendaho Graben, 273, 274, 276, **277–278**,
    285
Tendaho Sugar Plantation, 277
Termaber, 237–238

Tigray, 83–102
Tinish Fantale, Mt, 218
Tinish Sabober, tuff ring, 213, 217–218
Tis Isat Falls, 132, 304
Tiya, 233
Tosa Sucha, 191, 193
Tulu Dimtu, Mt (Western Ethiopia), 75, 76,
    80
Tulu Dimtu, Mt (Bale Mts), 159
Tulu Welel, Mt, 81–82, 118, 119
Turkana, Lake, 197
Turmi, 199

U
Urji volcano, 185, 190

V
Valley of Marvels, 164–165
Victoria, Lake, 174

W
Webi Gestro. *See* Gestro
Webi Shebeli. *See* Shebeli

Wachacha, 229, 230
Western Highlands, 19, 20, 22, 23,
    103–120, 121, 127, 131
White Nile, River, 57, 129, 133
Wonchi Mt, 119, 120
Woranso-Mille, 282
Wukro, xxxiv, 35

Y
Yabelo, 200, 202
Yangudi, Mt, 274
Yeha, 98
Yemen, 86, 106, 247
Yerer, Mt, 118, 230
Yubdo, 33, 74, 78

Z
Zagros, Mts, 172, 249
Zen Akwashita, lava tunnel, 125
Zula, Gulf of, 258, 262
Zuquala, Mt, 230
Zwai, Lake, 187

Index of Topics

A

Abreha and Atsbeha, 92

Adigrat Sandstone, **36**, 60, 61, 63, 66, 67, 69, 70, 85, 88, 167

Adola Belt, 198–199, **205–208**

Afar plume, **43**, 45, 296

Afar Triple Junction. *See* Triple junction

African Plate, 6, 8, 39–41, 45, 47, 172, 216, 249, 250, 295

African Rift System, **48–49**, 172, 174, 245

Afro-Arabian Dome, **43**, 59, 106, 131, 227, 296

Agula Shale, 87

Amazonite, 201

Amba, **87**, 95, 96

Amba Aradam Sandstone. *See* Upper Sandstone

Amole, 261

Anhydrite, 264

Antalo Limestone, **37–38**, 60, **61–63**, 66, 86, 167

Arabian-Nubian Shield, **30**, 70, 73, 85, 198, 205

Arabian Plate, 8, 49, 172, 247, 249, 301

Ardipithecus ramidus, 281, 282

Australopithecus Afarensis, 279–282

Australopithecus garhi, 281, 282

Australopithecus robustus, 203

Azania, **29**, 30, 31, 75

B

Birthplace of mankind, 271, **279–283**

Black smokers, 264

Blister, volcanic, 215–217

Bonga-Goba Lineament, 234

Brachiopods, 61, 64

Broadly Rifted Zone, **175**, 196, **197**, 205

Bruce, James, 128

C

Caldera, **182–184**, 185, 187, 190, 210–213, 254–257

Carlsberg Ridge, 39, 45

Caves, 125, **161–164**

    Cave churches, Lalibela, 143–150

    Cave paintings. *See* Rock art

Chert, 107, 108

Churches

    Lake Tana churches, 124, 128

    Lalibela rock churches, 135–151

    Tigray rock churches, 88–95

Cinder cones, 122, 124, 183, 193, 204, 209, 210, 212, 230, 234, 278

Cirque, 158, 159

Cleavage planes (slatey), 71

Columnar jointing, **59**, 62, 95, 107, 149, 222, 272

Copper, 77, 78

Corals, fossil, 258, 262

Cornwallis Harris, Major W., 212, 216, 289

Cross-bedding, 64

D

Danakil Microplate, 268

Debre Libanos Sandstone. *See* Upper Sandstone

Deinotherium, 108

© Springer International Publishing Switzerland 2016
F.M. Williams, *Understanding Ethiopia*, GeoGuide,
DOI 10.1007/978-3-319-02180-5

Diatomite, 190, 191
Dinkanesh. *See* Lucy
Dinosaurs, 66
Dolomite, 61–63, 65
Dunite, 76, **78**
Dyke, 93, 94, 106, 107, 169, 268

E
Earthquakes, **285–293**, 296
East African Orogen, **29**, 30, 31, 33, 34, 39, 75
East African Orogeny, **29**, 30, 47, 49, 70, 73, 75, 174, 205, 221
Eastern Rift, 173, **174**, 175
Edaga Arbi Tillite, 85
Enticho Sandstone, **85**, 88–90, 92, 96
Ethiopian Rift Valley, 19, 20, 49, **171–196**, **209–223**
Explosion crater. *See* Maar

F
Fairy chimneys. *See* Hoodoos
Fault, Definition, 8
Fiamme, 214
Fissures, **215–216**, **218**, 251, 253, 267–268, 278
Formation, Definition, 59
Fossil, 61, 64, 66, 67, 107, 108, 143, 144, 258, 262, 279, 281, 283
Fossil hominids. *See* Prehistoric sites
Fossil pollen. *See* Pollen, fossil
Fossil wood. *See* Wood, fossil
Fumaroles, 177, 185, 186, 251, 263, 278

G
Gabbro, 16, 76
Galla Lakes, 187–192
Gastropod, 61
Genesis, Book of, 103, 130
Geothermal fields, 185, 262–264, 278
Geyser, 262, 291
Ghion, 130

Gihon. *See* Ghion
Giulietti, Gueseppe, 245, 258
Glacial features (Bale Mts.), 158–160
Glacial features (Semien Mts.), 116
Glaciation, Palaeozoic, 35–36, 85
Glaciation, Pleistocene, 116, 158–160
Gneiss, **79**, 89, 156, 164, 165, 198–200, 201, 202, 204
Goha Tsion Formation, 60, 61, **63**, 65
Gold, 77, 78, **206–207**
Gold mines, 206–207
Gondwana, **28**, **29**, **30**, 33, **34**, 36, 39, 40
Graben, **195–196**, 197, 205, 246, 273, 275–278
Graben, marginal, 226, 232, 236, **238–241**
Granite, **16**, 29, 85, 164, 198, 199
Granite Intrusions, Afar, 269
Granite weathering, **164**, 198, 199, 202
Granitoid, 75
Gypsum, 37, 60, 61, **62–63**, 65, 146, 149, 167, 264

H
Hamanlei Limestone, 167
Headward erosion, **114**, 127, 131, 155, 159
Hildebrandt, Johann Maria, 256–257
Homo erectus, 203, 233, 282
Homo sapiens, 282
Hoodoos, 201
Horst, **195–196**, 197, 205, 246, 268
Hot-spot, **41–42**, 276, 298
Hotsprings, 119, 177, 184, 185, 189, 218, 220, 264, 278
Hyaloclastite rings, 258

I
Igneous rocks (Definition and Table of), **15–16**
Ignimbrite, 16, 106, 108, 111, 117, 128, 176, **182–184**, 187, 189, 194, 209, 211, 214, 218, 222, 233, 239, 241, 271, 279
Inselberg, 199

Intertrappean sediments, 107–108
Intrusion, **29**, 73, 75, 85, 93, 100, 169, 269
Inverted topography, 169, 170

K
Karrayu, people, 209
Karst, 86. *See also* Caves
King Solomon's Mine, 82

L
Lake sediments, 108, 182, 183, 190, 191, 209, 279
Lakes (*See* under localities index)
Lakes, Rift Valley, 187–193
Lalibela, King, 135
Lava, definition, 15
Lava fountains, 256, 257
Lava lake, 254, **256–258**
Lava tunnel, 124, 125
Limestone, 15, 37, 38, **61–63**, 86, 155, 161, 162, 167–169, 235, 242
Lucy (Dinkanesh), 279–280

M
Maar, **203–205**, 230–232, 234
Magma, **15**, 16, 29, 32, 75, 100, 169, 182, 184, 203, 223, 296
Main Ethiopian Rift. *See* Ethiopian Rift Valley
Main Ethiopian Rift, Formation of, 176
Mantle plume, 42
Marda Fault Zone, **165–166**, 168
Marginal graben. *See* Graben
Menelik II, Emperor, 98, 103
Mesa, **87**, 88. *See also* amba
Metamorphic rocks, **17**, 30, 73, 77, 197, 296
Metamorphism, Definition, 17
Mid-Atlantic Ridge, 40, 264, 298
Middle Awash palaeo-anthropological sites, 274, 282
Migmatite, **75**, 77, 156, 157
Mineral deposits, 30, 77. *See also* gold
Moraines, Glacial, 116, 158, 159

Mozambique Belt, **30**, 31, 49, 67, 73, 75, 198
Mozambique Ocean, 27–28, 30
Mugher Mudstone, 60, 65–66

N
Nuée ardente, 182, 184, 213

O
Obelisk, 98–101
Obsidian, 16, **185–186**, 267
Oil exploration, 167–168
Olivine, 78, 204, 205
Olivine nodules, 204
Opal, 108–110
Ophiolite, 76–77
Orogeny. *See* East African Orogeny
Oromo, people, 153, 159, 161, 187
Owen Fracture Zone, 40–41, 45

P
Paintings, cave. *See* Rock art
Paintings, church, 89, 91, 150
Panning for gold, **78–79**, 206, 207
Pastori, Tullio, 257
Pegmatite, **78–79**, 206, 207
Pele's Hair, 256
Phonolite, 96
Pitchstone, 16, **185**, 210, 253
Plate tectonics, 5–8
Platinum, 78
Plug, volcanic, 81–82, **96–98**, 109, **160**
Pollen, fossil, 64, 107, 283
Potash, 259
Potash, mining, 259
Precambrian basement, 35, 66, 69, 70, **88–89**, 164, 197, 221, 235, 268
Prehistoric sites, 161, 164, 203, 232–233, 279–283
Prester John, 103–104
Pumice, 16, 17, 182, 183, 185, 187, 211, 215, 267
Pyroxenite, 76

Q

Quarry, Axum, 100–101

R

Red Sea, 47, 49, 172, 174, 177, 247, 249, 250, 259, 270, 293, 296–298, 301
Red Series, 269–270
Rhyolite, 16, 108, 117, 118, 185, 210, 230, 239, 241, 253, 265, 269, 271, 279
Rift margin up-tilt, 154, **227**
Rift margins, **225–242**
Rift System. *See* African Rift System
Rift valley. *See* Ethiopian Rift Valley
Rifted basin, **36–37**, 64, 85, 124, 167
Rock art, 163–164
Rock churches, Lalibela, 135–151
Rock churches, Tigray, 88–95
Rodinia, 27–28

S

Salt castles, 264, 266
Salt mining, 259–261
Salt Plain, 258–262
Sapphire, 199, 202
Sea-floor spreading, **6**, 7, 9, 37, 39, 243, 249, 296–298
Sea-floor spreading axis, **6**, 47, 51, 171, 297
Sedimentary rocks, **15**, 22, 64, 83, 85, 117, 155, 268
Shear zone/shearing, **75–76**, 77, 205–206
Sheba Ridge, 45–47
Shield volcanoes, 43–44, 69–70, 111–117, 122, 157–161
Slate, 70–71
Somali, people, 153
Somali Plate, 46, 49, 226, 249, 250, 295
Spodumene, 207, 208
Spreading axis. *See* Sea-floor spreading axis
Spring sapping, 114
Stelae, 99, 233
Stone Age sites. *See* Prehistoric sites
Stratoid Series, **271–275**, 298

Striations, glacial. *See* glacial features
Syenite, 16, 98–100
Sylvite. *See* Potash

T

Tantalite, 207–208
Tantalum, 207–208
Termite mound, 199, 201
Tethys Sea, 40, 45, 47
Tillite, **35**, 85, 116
Tor, 199, 200
Trachyte, 16, 81, 96, 118, 157, 159, 185, 210, 230, 253, 267
Trap Series, 43–44
    Lalibela, 137
    Western Highlands, **106–111**, 117
    Blue Nile Gorge, **59–62**, 68
    Southeastern Highlands, 154–157
    Tekeze Gorge, 69–70
    Tigray, 83, 85, 87, 95, 98
    Western Ethiopia, 73
Triple junction, 49, 51, 245, 274, **276–277**, 296, 301
Tuff, 16, **137–142**, 145–147. *See also* Welded tuff
Tuff ring, 213, 217, 219

U

Ultramafic rocks, **76**, 78
Unconformity, 60, **62**, 66
Upper sandstone (Debre Libanos, Amba Aradam Sandstone), 38, 65, 66, 68, 87
U-shaped valley. *See* glacial features

V

Volcanic blisters. *See* Blister, volcanic
Volcanic plugs. *See* Plugs, volcanic
Volcanic ranges, 51, 246, **251–257**, 264–268, 278, 292, 298

W

Welded tuff, 16, 182, 215–217
West Tana Escarpment, **124**, **127**, 303
Western Escarpment, **235–241**, 262, 269, 270
Western Rift, 174
Wonji Fault Belt, **193–195**, 210, **220–223**, 276, 293
Wood, fossil, 64, 67, 143, 146

Y

Yerer-Tulu Welel Volcanotectonic Lineament, 105, **118–119**, 229

Z

Zircon, 32, 80, 81